数控车铣加工
高级（中英双语版）

郎卫珍 周 京 赵 慧 主 编
王 称 李成营 王 敏 张海涛 副主编

清华大学出版社
北京

内容简介

本书是"数控车铣加工"系列教材高级分册，内容涵盖数控加工工艺、编程基础和机床操作三部分。本书教学内容紧密对接教育部发布的"1+X"数控车铣加工职业技能等级标准，重点介绍数控车床、数控铣床和加工中心的编程指令、夹具、刀具、工艺流程、设备操作等专业知识，并且根据技能型人才培养需求，科学设计了典型教学案例及配套教学资源。通过渐进式的任务学习及训练，学生能够掌握数控编程及加工的基本方法、工艺常识和操作技能。

本书根据高职高专的教学特点，结合高职高专学生的实际学习能力和教学培养目标编写，可作为高职高专机械加工专业或其他相关专业的通用教材，也可作为成人教育院校的培训教材，还可供从事机械加工的工程技术人员参考。

本书封面贴有清华大学出版社防伪标签，无标签者不得销售。
版权所有，侵权必究。举报：010-62782989，beiqinquan@tup.tsinghua.edu.cn。

图书在版编目(CIP)数据

数控车铣加工：高级：汉、英/郎卫珍，周京，赵慧主编. —北京：清华大学出版社，2024.5
ISBN 978-7-302-66284-6

Ⅰ. ①数… Ⅱ. ①郎… ②周… ③赵… Ⅲ. ①数控机床－车床－加工工艺－职业技能－鉴定－教材－汉、英 ②数控机床－铣床－加工工艺－职业技能－鉴定－教材－汉、英 Ⅳ. ①TG519.1 ②TG547

中国国家版本馆 CIP 数据核字(2024)第 098087 号

责任编辑：王　芳　薛　阳
封面设计：刘　键
责任校对：王勤勤
责任印制：杨　艳

出版发行：清华大学出版社
　　　　网　　址：https://www.tup.com.cn，https://www.wqxuetang.com
　　　　地　　址：北京清华大学学研大厦 A 座　　邮　　编：100084
　　　　社 总 机：010-83470000　　邮　　购：010-62786544
　　　　投稿与读者服务：010-62776969，c-service@tup.tsinghua.edu.cn
　　　　质量反馈：010-62772015，zhiliang@tup.tsinghua.edu.cn
　　　　课件下载：https://www.tup.com.cn，010-83470236
印 装 者：三河市君旺印务有限公司
经　　销：全国新华书店
开　　本：185mm×260mm　　印　张：13　　字　数：318 千字
版　　次：2024 年 5 月第 1 版　　印　次：2024 年 5 月第 1 次印刷
印　　数：1～1500
定　　价：59.00 元

产品编号：098632-01

前　　言

随着自动化、数字化、网络化、智能化技术的快速发展及广泛应用,制造业的人才需求发生了很大变化,其需求对象由某一个领域单一技术的技能人才转变为"通才＋专才"复合型技术的技能人才。为进一步落实中国共产党第十九次全国代表大会提出的深化"产教融合"的重大任务,国务院在发布的《国家职业教育改革实施方案》中明确提出在职业院校、应用型本科高校启动"学历证书＋若干职业技能等级证书"制度(即"1＋X"职业技能等级证书制度)试点工作,明确了开展深度"产教融合""双元"育人的具体指导政策与要求。其中,"1＋X"职业技能等级证书制度是统筹考虑、全盘谋划职业教育发展、推动企业深度参与协同育人和深化复合型技术技能人才培养培训而做出的重大制度设计。

本书是"1＋X"职业技能等级证书(数控车铣加工)系列教材之一,是根据教育部数控技能型紧缺人才培养培训方案的指导思想,以及数控车铣加工职业技能等级证书的标准要求,结合当前数控技术的发展及教学规律编写而成的。本系列教材以数控车铣加工职业技能等级证书考核样题为基础,选用国内多种通用的CAD/CAM软件,从数控车和数控铣产品加工的典型任务入手,通过讲解工程图样及工艺文件,编制零件的数控加工工艺和加工程序,特别是针对数控车铣综合加工工艺进行案例分析,使学习者能够掌握数控机床加工编程,完成定位及联动加工、检测,并控制产品的加工精度、对数控机床的精度进行检验及排除数控机床的一般故障等技能。

目前,已经出版的数控车铣机床高职教材存在教学思路老套、教学内容更新慢、配套资源形式单一,以及教材贴合度差等弊端,且缺少思政资源,不满足当前的立体化、多媒体化、电子化、思政元素深度化等教学要求。针对这些现状,本书以最新的车铣设备作为编写硬件,采用EPIP教学模式,突出实践技能训练,深度递进式写作方法,配套电子资源包括课程标准、PPT、微课、动画、教学录像等,是数控车铣教学的"交钥匙工程"项目,在图书市场上还未见到该类型出版物。本书出版后,在中职、高职学校及数控设备加工人员培训等市场上,会有很大的用户群体。

本书共分6个项目,项目一由郎卫珍编写,项目二由周京编写,项目三由赵慧编写,项目四由王称编写,项目五由李成营编写,项目六由王敏编写,张海涛、王丹阳参与了本书的翻译工作。

由于编者水平有限,书中难免会有不足,恳请使用本书的师生和读者批评指正。

编　者

2023年11月

Preface

Luban Workshop is a well-known brand for cultural exchanges between China and foreign countries, which is first practiced in Tianjin under the strong support and guidance of the Ministry of Education of China. It is committed to cultivating technical and skilled talents who are familiar with Chinese technology, understand Chinese processing and recognize Chinese products for the cooperative countries. It is a landmark achievement of the National Modern Vocational Education Reform and Innovation Demonstration Zone and a major innovation in the international development of China's vocational education. Since the first Luban Workshop was established in Thailand in 2016, China has successively carried out cooperation in countries along "the Belt and Road" to build a new stage for Sino-foreign vocational education cooperation. In December 2018, Tianjin Light Industry Vocational Technical College, together with Tianjin Transportation Technical College, cooperated with Egypt's Ainshams University and Cairo Advanced Maintenance Technology School to jointly build Luban Workshop, in which the CNC machining technology is one of the key majors at the secondary vocational education level of the Luban workshop jointly built by China and Egypt. In order to cooperate with the theoretical and practical teaching of the "Luban Workshop" in Egypt, carry out exchanges and cooperation, improve the international influence of China's vocational education, innovate the international cooperation mode of vocational colleges, and export the excellent resources of China's vocational education, the research group has prepared this book.

The Engineering Practice Innovation Project (EPIP) teaching mode is the core content of Luban Workshop. It integrates theoretical teaching and practical teaching, and forms and develops students' comprehensive professional ability and innovation ability in real work situations. This textbook is based on the EPIP teaching mode, taking the "Luban Workshop" CNC processing equipment as the carrier, combining with the current development of CNC turning and milling technology, taking the actual engineering project as the guide, and taking the practical application as the guide, while paying attention to the basic theory education, highlighting the practical skill training, and cultivating high-level skilled talents with excellent scientific research ability and problem-solving ability.

This book, as the advanced engineering volume of the series of "CNC Turning and Milling", takes the suction pump model as an example, and uses CAXA CAM CNC lathe and CAXA CAM manufacturing engineer programming software to gradually explain the process design, software operation and post-processing code generation from easy to difficult. The suction pump device mainly consists of 9 typical parts. This book takes the first 6 typical parts of the suction pump model as the carrier and takes six typical projects as the main line. It focuses on the professional knowledge and operating skills required for the turning programming and processing of the one-way valve, pump body and crank as well as the milling programming and processing of the support seat, base frame and crank connecting rod.

This book adopts a deep and progressive writing method, and tries to make it profound but easily understood, highlighting the characteristics of higher vocational education. Supporting electronic resources include curriculum standards, PPT, micro-class, animation, teaching videos, etc., which are convenient for teaching and suitable for engineering students in higher vocational colleges. This textbook is written in both Chinese and English, and is also suitable for the students of Luban Workshop in Egypt who major in CNC machining technology.

This book contains 6 projects. Project 1 is written by Lang Weizhen. Project 2 is written by Zhou Jing. Project 3 is written by Zhao Hui. Project 4 is written by Wang Chen. Project 5 is written by Li Chengying. Project 6 is written by Wang Min. Zhang Haitao and Wang Danyang participate in the translation of this book.

Due to the limited knowledge of editors, some mistakes and errors are unavoidable, welcome all readers to give kind feedback.

Editor

Nov. 2023

目　　录

项目引导 ··· 1

项目一　单向阀车削编程加工训练 ·· 2

 任务一　学习关键知识点 ··· 3
  1.1　数控车软件基本功能介绍 ·· 3
  1.2　基本操作介绍 ·· 5
   1.2.1　窗口布局 ·· 5
   1.2.2　鼠标和键盘命令 ·· 6
  1.3　数控车加工基本概念介绍 ·· 6
   1.3.1　两轴加工 ·· 6
   1.3.2　轮廓 ·· 7
   1.3.3　毛坯轮廓 ·· 7
  1.4　数控车软件刀库设置 ·· 7
   1.4.1　轮廓车刀 ·· 8
   1.4.2　切槽车刀 ·· 9
   1.4.3　螺纹车刀 ·· 9
   1.4.4　钻头 ·· 10
  1.5　数控车软件后置设置 ·· 11
   1.5.1　"通常"选项卡 ·· 12
   1.5.2　"运动"选项卡 ·· 12
   1.5.3　"主轴"选项卡 ·· 13
   1.5.4　"地址"选项卡 ·· 14
   1.5.5　"关联"选项卡 ·· 15
   1.5.6　"程序"选项卡 ·· 15
   1.5.7　"车削"选项卡 ·· 16
   1.5.8　"机床"选项卡 ·· 17
  1.6　基本指令 ·· 18
   1.6.1　"车削粗加工（创建）"对话框 ·· 18
   1.6.2　"加工参数"选项卡 ·· 19
   1.6.3　"进退刀方式"选项卡 ··· 20
   1.6.4　"切削用量"选项卡 ·· 21
   1.6.5　"轮廓车刀"选项卡 ·· 22
 任务二　工艺准备 ·· 22

  1.7 零件图分析 ·· 22
  1.8 工艺设计 ··· 22
 任务三 软件编程训练 ··· 23
  1.9 编程练习 ··· 23
   1.9.1 零件造型 ··· 23
   1.9.2 加工刀具轨迹 ·· 24
   1.9.3 反向装夹造型 ·· 25
   1.9.4 反向装夹加工刀具轨迹 ·· 25
 项目总结 ·· 25
 课后习题 ·· 26

项目二 泵体车削编程加工训练 ·· 28

 任务一 学习关键知识点 ··· 29
  2.1 "车削槽加工（创建）"对话框 ······································ 29
   2.1.1 "加工参数"选项卡 ··· 30
   2.1.2 "切削用量"选项卡 ··· 31
   2.1.3 "切槽车刀"选项卡 ··· 31
   2.1.4 车削槽加工实例 ·· 32
  2.2 "车螺纹加工（创建）"对话框 ······································ 33
   2.2.1 "螺纹参数"选项卡 ··· 34
   2.2.2 "加工参数"选项卡 ··· 34
   2.2.3 "进退刀方式"选项卡 ·· 36
   2.2.4 "切削用量"选项卡 ··· 36
   2.2.5 "螺纹车刀"选项卡 ··· 37
 任务二 工艺准备 ·· 38
  2.3 零件图分析 ·· 38
  2.4 工艺设计 ··· 38
 任务三 软件编程训练 ··· 40
  2.5 编程练习 ··· 40
   2.5.1 零件造型 ··· 40
   2.5.2 加工刀具轨迹 ·· 40
   2.5.3 反向装夹造型 ·· 41
   2.5.4 反向装夹加工刀具轨迹 ·· 41
 项目总结 ·· 42
 课后习题 ·· 42

项目三 曲柄车削编程加工训练 ·· 45

 任务一 学习关键知识点 ··· 46
  3.1 四爪卡盘简介 ·· 46
   3.1.1 介绍 ·· 46

       3.1.2　适用范围 …………………………………………………… 47
       3.1.3　用途 ……………………………………………………… 47
     3.2　车削偏心零件的加工方法 ………………………………………… 47
       3.2.1　利用三爪卡盘装夹 …………………………………………… 47
       3.2.2　利用四爪单动卡盘装夹 ……………………………………… 48
       3.2.3　偏心轮车夹具 ………………………………………………… 48
   任务二　工艺准备 …………………………………………………………… 49
     3.3　零件图分析 ………………………………………………………… 49
     3.4　工艺设计 …………………………………………………………… 49
   任务三　软件编程训练 ……………………………………………………… 51
     3.5　编程练习 …………………………………………………………… 51
       3.5.1　零件造型 ……………………………………………………… 51
       3.5.2　加工刀具轨迹 ………………………………………………… 51
       3.5.3　反向装夹调偏心后造型及加工刀具轨迹 …………………… 52
   项目总结 ……………………………………………………………………… 53
   课后习题 ……………………………………………………………………… 53

项目四　支撑座铣削编程加工训练 ………………………………………… 56

   任务一　学习关键知识点 …………………………………………………… 57
     4.1　制造工程师软件基本功能介绍 …………………………………… 57
     4.2　基本操作介绍 ……………………………………………………… 57
     4.3　制造工程师加工基本概念介绍 …………………………………… 57
       4.3.1　造型 …………………………………………………………… 57
       4.3.2　编程助手 ……………………………………………………… 58
       4.3.3　等高线加工 …………………………………………………… 58
   任务二　工艺准备 …………………………………………………………… 60
     4.4　零件图分析 ………………………………………………………… 60
     4.5　工艺设计 …………………………………………………………… 61
   任务三　软件编程训练 ……………………………………………………… 63
     4.6　编程练习 …………………………………………………………… 63
       4.6.1　创建加工坐标系及毛坯 ……………………………………… 63
       4.6.2　加工刀具轨迹 ………………………………………………… 64
       4.6.3　反向装夹加工刀具轨迹 ……………………………………… 65
       4.6.4　侧向装夹孔加工刀具轨迹 …………………………………… 65
       4.6.5　底部不规则槽加工刀具轨迹 ………………………………… 66
   项目总结 ……………………………………………………………………… 66
   课后习题 ……………………………………………………………………… 67

项目五　底座铣削编程加工训练 … 69

任务一　学习关键知识点 … 70
5.1　螺纹铣削加工 … 70
5.1.1　螺旋铣削内孔 … 70
5.1.2　单刃螺纹铣刀加工螺纹 … 70
5.1.3　多刃螺纹铣刀加工螺纹 … 70
5.2　曲面加工 … 71
5.2.1　三轴曲面加工的意义 … 71
5.2.2　三轴曲面加工方法 … 72

任务二　工艺准备 … 72
5.3　零件图分析 … 72
5.4　工艺设计 … 73

任务三　软件编程训练 … 75
5.5　编程练习 … 75
5.5.1　创建毛坯及加工坐标系 … 75
5.5.2　加工刀具轨迹 … 75
5.5.3　反向装夹加工刀具轨迹 … 77

项目总结 … 77
课后习题 … 78

项目六　曲柄连杆铣削编程加工训练 … 80

任务一　学习关键知识点 … 80
6.1　切槽铣刀 … 80
6.2　夹具及辅助支撑 … 82

任务二　工艺准备 … 82
6.3　零件图分析 … 82
6.4　工艺设计 … 83

任务三　软件编程训练 … 85
6.5　编程练习 … 85
6.5.1　创建毛坯及加工坐标系 … 85
6.5.2　加工刀具轨迹 … 85
6.5.3　反向装夹加工刀具轨迹 … 86

项目总结 … 87
课后习题 … 87

Contents

Projects Guidance ……………………………………………………………… 89

Project 1 Programming and Machining Training for One-way Valve Turning ……… 91

  Task 1 Learn Key Knowledge Points ……………………………………………… 92
    1.1 Introduction to basic functions of the CNC lathe ……………………… 92
    1.2 Introduction to basic operations …………………………………………… 95
      1.2.1 Window layout ………………………………………………………… 95
      1.2.2 Mouse and keyboard commands ……………………………………… 96
    1.3 Introduction to basic concepts of CNC turning …………………………… 96
      1.3.1 Two-axis machining …………………………………………………… 96
      1.3.2 Contour ………………………………………………………………… 97
      1.3.3 Blank contour ………………………………………………………… 97
    1.4 Tool magazine settings of the CNC lathe software ……………………… 97
      1.4.1 Contour turning tools ………………………………………………… 98
      1.4.2 Grooving turning tools ………………………………………………… 99
      1.4.3 Thread turning tools ………………………………………………… 100
      1.4.4 Drilling tools ………………………………………………………… 101
    1.5 Post-setting of the CNC lathe software …………………………………… 102
      1.5.1 General tab …………………………………………………………… 103
      1.5.2 Motion tab …………………………………………………………… 103
      1.5.3 Spindle tab …………………………………………………………… 104
      1.5.4 Address tab …………………………………………………………… 105
      1.5.5 Association tab ……………………………………………………… 106
      1.5.6 Program tab …………………………………………………………… 106
      1.5.7 Turning tab …………………………………………………………… 108
      1.5.8 Machine tool tab …………………………………………………… 109
    1.6 Basic instructions …………………………………………………………… 109
      1.6.1 Rough turning(create)dialog box …………………………………… 109
      1.6.2 Machining parameters tab …………………………………………… 110
      1.6.3 Feed and retract mode ……………………………………………… 112
      1.6.4 Cutting dosage tab …………………………………………………… 114
      1.6.5 Contour turning tools ………………………………………………… 115
  Task 2 Technological Preparation ………………………………………………… 115

  1.7 Part drawing analysis ················· 115
  1.8 Technological design ················· 115
 Task 3 Software Programming Practice ················· 117
  1.9 Programming practice ················· 117
   1.9.1 Part modelling ················· 117
   1.9.2 Tool path ················· 117
   1.9.3 Reverse clamping for modelling ················· 118
   1.9.4 Reverse clamping for roughing tool path ················· 118
 Project Summary ················· 119
 Exercises After Class ················· 119

Project 2 Programming and Machining Training for Pump Body Turning ················· 121

 Task 1 Learn Key Knowledge Points ················· 122
  2.1 Groove turning(create)dialog box ················· 122
   2.1.1 Machining parameters tab ················· 123
   2.1.2 Cutting dosage tab ················· 124
   2.1.3 Groove turning tools tab ················· 125
   2.1.4 Machining example of turning a groove ················· 126
  2.2 Thread turning(create)dialog box ················· 127
   2.2.1 Thread parameters tab ················· 128
   2.2.2 Machining parameters tab ················· 128
   2.2.3 Feed and retract mode tab ················· 130
   2.2.4 Cutting dosage tab ················· 131
   2.2.5 Threading tool tab ················· 131
 Task 2 Technological Preparation ················· 132
  2.3 Part drawing analysis ················· 132
  2.4 Technological design ················· 132
 Task 3 Software Programming Practice ················· 135
  2.5 Programming practice ················· 135
   2.5.1 Part model ················· 135
   2.5.2 Tool path ················· 136
   2.5.3 Reverse clamping modelling ················· 136
   2.5.4 Reverse clamping machining tool path ················· 137
 Project Summary ················· 138
 Exercises After Class ················· 138

Project 3 Programming and Machining Training for Crank Turning ················· 141

 Task 1 Learn Key Knowledge Points ················· 142
  3.1 Introduction to the four-jaw chuck ················· 142
   3.1.1 Introduction ················· 142

　　　　3.1.2　Scope of application ······ 143
　　　　3.1.3　Purpose ······ 143
　　3.2　Method of turning eccentric workpiece ······ 144
　　　　3.2.1　Clamping with a three-jaw chuck ······ 144
　　　　3.2.2　Clamping with a four-jaw single-action chuck ······ 145
　　　　3.2.3　Eccentric wheel lathe fixture ······ 145
　Task 2　Technological Preparation ······ 146
　　3.3　Part drawing analysis ······ 146
　　3.4　Technological design ······ 146
　Task 3　Software Programming Practice ······ 148
　　3.5　Programming practice ······ 148
　　　　3.5.1　Drawing modeling ······ 148
　　　　3.5.2　Machining tool path ······ 149
　　　　3.5.3　Modeling and machining tool path after reverse clamping and eccentric adjustment ······ 149
　Project Summary ······ 150
　Exercises After Class ······ 151

Project 4　Programming and Machining Training for Supporting Seat Milling ······ 154

　Task 1　Learn Key Knowledge Points ······ 155
　　4.1　Introduction to basic functions of Manufacturing Engineer software ······ 155
　　4.2　Introduction to basic operations ······ 155
　　4.3　Introduction to basic concepts of Manufacturing Engineer ······ 155
　　　　4.3.1　Modeling ······ 155
　　　　4.3.2　Programming assistant ······ 156
　　　　4.3.3　Contour machining ······ 157
　Task 2　Technological Preparation ······ 159
　　4.4　Part drawing analysis ······ 159
　　4.5　Technological design ······ 160
　Task 3　Software Programming Practice ······ 163
　　4.6　Programming practice ······ 163
　　　　4.6.1　Create machining coordinate system and blank ······ 163
　　　　4.6.2　Machining tool path ······ 164
　　　　4.6.3　Machining tool path of reverse clamping ······ 165
　　　　4.6.4　Hole machining tool path of lateral clamping ······ 165
　　　　4.6.5　Machining tool path of bottom irregular groove ······ 166
　Project Summary ······ 166
　Exercises After Class ······ 167

Project 5　Programming and Machining Training for Base Frame Milling ······ 170

　Task 1　Learn Key Knowledge Points ······ 171

5.1 Thread milling ……………………………………………………………… 171
 5.1.1 Spiral milling of an inner hole ……………………………………… 171
 5.1.2 Machining thread with single-edge thread milling cutter …… 171
 5.1.3 Machining thread with multi-edge thread milling cutter …… 172
5.2 Surface machining ……………………………………………………… 172
 5.2.1 The significance of three-axis surface machining …………… 173
 5.2.2 Three-axis surface machining method ……………………… 173
Task 2 Technological Preparation ……………………………………………… 174
5.3 Part drawing analysis …………………………………………………… 174
5.4 Technological design …………………………………………………… 175
Task 3 Software Programming Practice ……………………………………… 177
5.5 Programming practice ………………………………………………… 177
 5.5.1 Creating blank and machining coordinate system ………… 177
 5.5.2 Machining tool path ……………………………………………… 178
 5.5.3 Reverse clamping machining tool path ……………………… 179
Project Summary …………………………………………………………………… 180
Exercises After Class ……………………………………………………………… 180

Project 6 Programming and Machining Training for Crank Connecting Rod Milling …………………………………………………………… 183

Task 1 Learn Key Knowledge Points …………………………………………… 183
6.1 Groove milling cutters ………………………………………………… 183
6.2 Fixture and auxiliary support ………………………………………… 185
Task 2 Technological Preparation ……………………………………………… 186
6.3 Part drawing analysis …………………………………………………… 186
6.4 Technological design …………………………………………………… 187
Task 3 Software Programming Practice ……………………………………… 189
6.5 Programming practice ………………………………………………… 189
 6.5.1 Creating blank and machining coordinate system ………… 189
 6.5.2 Tool path ………………………………………………………… 189
 6.5.3 Reverse clamping tool path …………………………………… 191
Project Summary …………………………………………………………………… 191
Exercises After Class ……………………………………………………………… 192

项目引导

数控编程是数控加工中的一项重要工作,从零件图纸到获得合格数控加工程序的过程称为数控编程。本书以吸水泵(见图 0-1)模型为实例,使用 CAXA CAM 数控车(后文简称数控车)与 CAXA CAM 制造工程师(后文简称制造工程师)编程软件,由易到难讲解工艺编制、软件操作及后置处理代码生成。

图 0-1　吸水泵

注意:编程软件版本为 2020 版,后文不再进行说明。此外,默认在学习本书之前,读者已经具备制图能力。

如图 0-2 所示,吸水泵主要包括 9 个典型零件,分别涉及数控车、数控铣和复合加工方案。以前六个零件为主线,逐一展开轴、孔、螺纹、槽、型腔等零件特征的软件编程方法。每个工作任务都包含零件加工所需专业知识和技能的学习,如设备、夹具、刀具、基本指令和机床操作等。根据数控加工的基本流程,进行有针对性的学习和训练,以达到熟练掌握数控车削和数控铣削加工工艺设计、数控编程和机床操作的目的(复合加工作为最后练习使用,不在本书中赘述)。

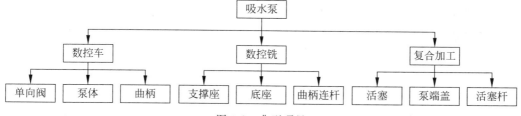

图 0-2　典型项目

项目一　单向阀车削编程加工训练

➢ **思维导图**

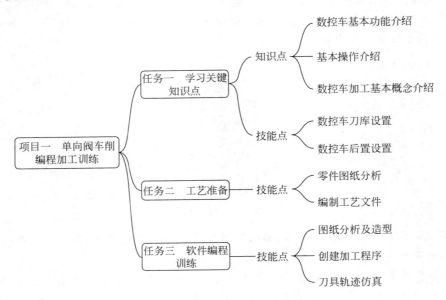

➢ **学习目标**

知识目标

(1) 了解 CAXA CAM 数控车软件窗口。
(2) 了解图纸在数控车软件中的图素处理。
(3) 理解加工参数选择和刀具轨迹仿真检验。

能力目标

(1) 对图纸进行分析,确定需要加工的部分。
(2) 利用图形软件对加工部分造型。
(3) 根据加工条件,选择合适的加工参数生成刀具轨迹。
(4) 刀具轨迹仿真检验。
(5) 配置好机床,生成 G 代码传输给机床加工。

素养目标

(1) 培养学生的学习热情。
(2) 培养学生的动手能力。

（3）建立学生的独立思考能力。

▶任务引入

根据零件图（见图1-1）要求，制定加工工艺，使用数控车软件编写加工程序，并完成单向阀零件的加工。该零件作为典型的阶梯轴类零件，材料为45钢，要求表面光整无划伤，不得用砂纸或锉刀修整零件表面。

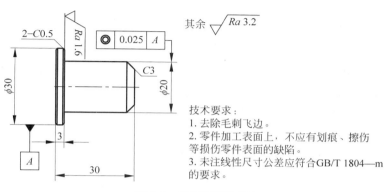

技术要求：
1. 去除毛刺飞边。
2. 零件加工表面上，不应有划痕、擦伤等损伤零件表面的缺陷。
3. 未注线性尺寸公差应符合GB/T 1804—m的要求。

图1-1 单向阀零件图

任务一 学习关键知识点

1.1 数控车软件基本功能介绍

数控车软件是在全新的数控加工平台上开发的数控车床加工编程和二维图形设计软件。数控车软件具有CAD软件的强大绘图功能和完善的外部数据接口，可以绘制任意复杂的图形，可通过DXF、IGES等数据接口与其他系统交换数据。数控车软件提供了功能强大、使用简洁的轨迹生成手段，可按加工要求生成各种复杂图形的刀具轨迹。通用的后置处理模块使数控车软件可以满足各种机床的代码格式，可以输出G代码，并对生成的NC代码进行校验及加工仿真。同时，也提供对NC代码文件的反读和刀具轨迹的校验、仿真功能。

2020版数控车软件基于二维平台CAXA电子图板打造，在操作界面和操作风格上都有变化。图1-2为2020版数控车软件菜单栏。

图1-2 2020版数控车软件菜单栏

根据在车削加工中常用的粗加工、精加工、切槽加工、螺纹加工等工序，在2020版数控车软件中开发出了专业的"车削粗加工"、"车削精加工"、"车削槽加工"和3种不同的螺纹加工功能（见图1-3）。

除了标准的二轴加工的功能以外，还针对带C轴的车削中心开发了等截面加工、径向和端面钻孔、埋入式和开放式键槽加工等功能（见图1-4）。

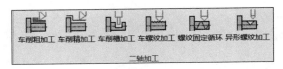

图 1-3 "二轴加工"选项组

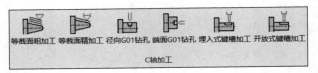

图 1-4 "C 轴加工"选项组

使用这些功能可以很方便地生成刀具轨迹,通过仿真功能可以校验刀具轨迹的正确性。在校验完成后,通过后置处理(见图 1-5)将轨迹转换成机床所需的 NC 代码文件。在数控车软件中,通过通信端口可以把代码文件传输给机床加工使用。系统已经内置一些通用后置处理文件,如 FANUC、SIEMENS 等数控系统后置处理。

图 1-5 "后置处理"选项组

(1) 开放的"后置设置"功能,用户可根据企业的机床自定义后置处理,允许根据特种机床自定义代码,自动生成符合特种机床的代码文件,用于加工。

(2) 支持小内存机床系统加工,支持自动将大程序分段输出功能。

(3) 根据数控系统要求是否输出行号,行号是否自动填写,编程方式可以选择增量坐标或绝对坐标方式编程。

(4) 坐标输出格式可以定义到小数及整数位数。

(5) 圆弧输出方式是用 I、G、K 及 R 代码各自的含义设定。

管理树栏(见图 1-6)以树形图的形式,直观地展示了当前文档的刀具、轨迹、代码等信

图 1-6 管理树栏

息,并提供了很多树上的操作功能,便于用户执行各项与数控车相关的命令。管理树栏框体默认位于绘图区的左侧,用户可以自由拖动它到喜欢的位置,也可以将其隐藏起来。管理树栏有一个"加工"总节点,总节点下有"刀库""轨迹""代码"三个子节点,分别用于显示和管理刀具信息、轨迹信息和 G 代码信息。

一体化刀库管理功能,包括轮廓车刀、切槽车刀、螺纹车刀、钻头多种刀具类型的管理。便于用户从刀库中获取刀具信息和对刀库进行维护。图 1-7 所示为刀库管理"创建刀具"对话框。

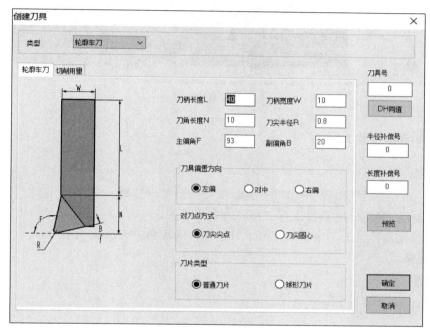

图 1-7　刀库管理"创建刀具"对话框

在 2020 版数控车软件中,可直接通过利用"创建刀具"对话框,或者在"菜单"栏选择"创建刀具"命令就可以很方便地创建一把刀具。而且,将这把刀具所用到的切削用量和几何参数进行关联,在加工过程中调用这把刀具的同时,就可以调用这把刀的切削要素,简化刀具在切削用量上重新设定参数的时间。

1.2　基本操作介绍

1.2.1　窗口布局

数控车软件的窗口布局如图 1-8 所示。

(1) 工具栏:所有的功能命令可以在选项卡区域进行查找。

(2) 管理树栏:所有的刀具、数控车刀具轨迹、G 代码信息都会被记录并显示在管理树栏上。

(3) 状态栏:选项卡功能运行选项及操作命令提示。

(4) 绘图区:支持多浏览,可在不同图纸间随意切换编辑。

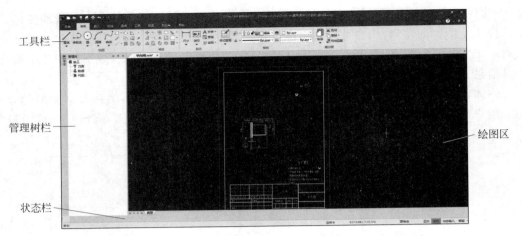

图 1-8 数控车软件的窗口布局

1.2.2 鼠标和键盘命令

1. 鼠标

鼠标左键主要担任拾取和确认的功能,可以通过单击选择一个功能、拾取坐标点或拾取元素等。鼠标右键主要用于确认当前命令结束或返回上一个命令,也可在特定区域内调出右键菜单,简化操作。上下滚动滚轮可以对绘图区的视角进行放大和缩小,按下滚轮可对绘图区进行平移。

2. 键盘

常用快捷键及定义方式如表 1-1 所示。

表 1-1 常用快捷键及定义方式

快捷键	定义	快捷键	定义	快捷键	定义
F1	帮助文件	F6	捕捉方式切换	F9	窗口切换
F3	显示全部	F7	三维视图导航开关	Delete	删除
F5	坐标系切换	F8	正交模式开关	Ctrl+P	打印

其余快捷键与常用软件一致,如复制的快捷键为 Ctrl+C 等。

1.3 数控车加工基本概念介绍

利用数控车软件实现加工的过程首先是对图纸进行分析,确定需要数控加工的部分;利用图形软件对需要加工的部分造型;根据加工条件,选择合适的参数生成刀具轨迹(包括粗加工、半精加工、精加工轨迹);刀具轨迹仿真检验;配置好机床,生成 G 代码传输给机床加工。

1.3.1 两轴加工

在数控车软件中,机床坐标系的 Z 轴即绝对坐标系的 X 轴,平面图形均投影到绝对坐标系的 XOY 面,如图 1-9 所示。

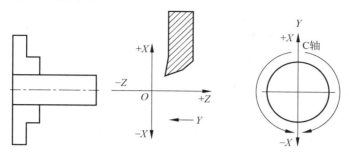

图 1-9　两轴加工

在一般情况下,需要建立一个加工坐标系,与机床坐标系形成一定的关系。

1.3.2　轮廓

数控车软件定义轮廓是一系列首尾相接的曲线集合,如图 1-10 所示。

1.3.3　毛坯轮廓

针对粗加工,在加工时需要定义加工范围,即毛坯轮廓,如图 1-11 所示。

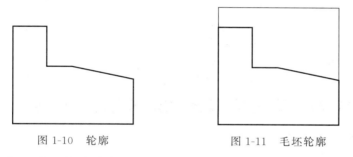

图 1-10　轮廓　　　　　　　图 1-11　毛坯轮廓

在进行数控编程,交互指定待加工的图形时,常常需要用户指定毛坯的轮廓,用来界定被加工的表面或被加工的毛坯本身。

注意:如果毛坯轮廓是用来界定被加工表面的,则要求指定的轮廓是闭合的;如果加工的是毛坯轮廓本身,则毛坯轮廓可以不闭合。

1.4　数控车软件刀库设置

在使用数控车软件进行加工前,需要对刀具、数控系统和机床进行设置,它们将直接影响到加工轨迹和生成的 G 代码。本章将对这些设置内容进行详细介绍。

刀库管理功能定义、确定刀具的相关数据,以便于用户从刀库中获取刀具信息,并对刀库进行维护。刀库管理功能包括轮廓车刀、切槽车刀、螺纹车刀、钻头 4 种刀具类型的管理。

操作方法如下。

(1) 在菜单栏中单击"数控车"标签,再单击"创建刀具"按钮,弹出"创建刀具"对话框,用户可按自己的需要添加新的刀具。新创建的刀具列表会显示在绘图区左侧的"管理树"→"刀库"节点下。

(2) 双击"刀库"节点下的"刀具"节点,可弹出"编辑刀具"对话框,用来改变刀具参数。

(3) 在"刀库"节点右击后,在弹出的快捷菜单中单击"导出刀具"命令,可以将所有刀具的信息保存到一个文件中。

(4) 在"刀库"节点右击后,在弹出的快捷菜单中单击"导入刀具"命令,可以将保存到文件中的刀具信息全部读入文档中,并添加到"刀库"节点下。

(5) 需要指出的是,刀库中的各种刀具只是对同一类刀具的抽象描述,并非符合国标或其他标准的详细刀库。因此,只列出了对轨迹生成有影响的部分参数,其他与具体加工工艺相关的刀具参数并未列出。例如,将各种外轮廓、内轮廓、端面粗、精车刀均归为轮廓车刀,对轨迹生成没有影响。

1.4.1 轮廓车刀

"轮廓车刀"选项卡如图 1-12 所示,需要配置的参数具体如下。

(1) "刀具号":刀具的系列号,用于后置处理的自动换刀指令。刀具号唯一,且对应机床的刀库。

(2) "半径补偿号":刀具半径补偿值的序列号,其值对应于机床的数据库。

(3) "刀柄长度":刀具可夹持段的长度。

(4) "刀柄宽度":刀具可夹持段的宽度。

(5) "刀角长度":刀具可切削段的长度。

(6) "刀尖半径":刀尖部分用于切削的圆弧半径。

(7) "主偏角":刀具主切削刃与工件旋转轴之间的夹角。

(8) "副偏角":刀具副切削刃与工件旋转轴之间的夹角。

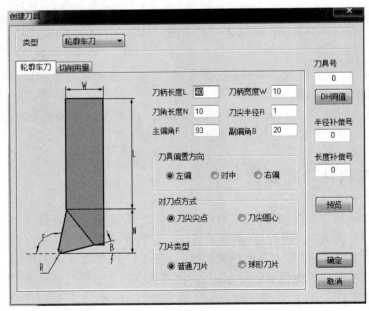

图 1-12 "轮廓车刀"选项卡

May all your wishes come true

清华大学出版社

如果知识是通向未来的大门，
我们愿意为你打造一把打开这扇门的钥匙！

https://www.shuimushui.com/

图书详情 | 配套资源 | 课程视频 | 会议资讯 | 图书出版

扬帆起航

1.4.2 切槽车刀

"切槽车刀"选项卡如图1-13所示,需要配置的参数具体如下。

(1)"刀具号":刀具的系列号,用于后置处理的自动换刀指令。刀具号唯一,且对应机床的刀库。

(2)"半径补偿号":刀具半径补偿值的序列号,其值对应于机床的数据库。

(3)"长度补偿号":刀具长度补偿值的序列号,其值对应于机床的数据库。

(4)刀具长度:刀具可夹持段的长度。

(5)刀具宽度:刀具可夹持段的宽度。

(6)刀刃宽度:刀具可切削段的长度。

(7)刀尖半径:刀尖部分用于切削的圆弧半径。

主偏角:刀具主切削刃与工件旋转轴之间的夹角。

副偏角:刀具副切削刃与工件旋转轴之间的夹角。

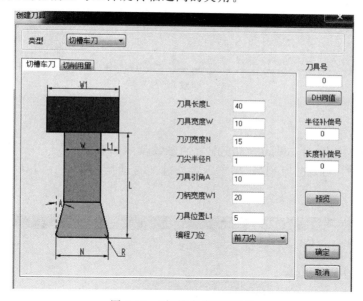

图1-13 "切槽车刀"选项卡

1.4.3 螺纹车刀

"螺纹车刀"选项卡如图1-14所示,需要配置的参数具体如下。

(1)"刀具号":刀具的系列号,用于后置处理的自动换刀指令。刀具号唯一,且对应机床的刀库。

(2)"半径补偿号":刀具半径补偿值的序列号,其值对应于机床的数据库。

(3)"刀柄长度":刀具可夹持段的长度。

(4)"刀柄宽度":刀具可夹持段的宽度。

(5)"刀刃长度":刀具切削刃顶部的宽度。

(6)"刀尖宽度":螺纹齿槽宽度。

(7)"刀具角度":刀具切削段两侧边与垂直于切削方向的夹角,该角度决定了车削出的螺纹的牙型角。

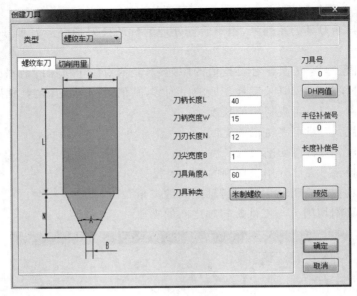

图 1-14 "螺纹车刀"选项卡

1.4.4 钻头

"钻头"选项卡如图 1-15 所示,需要配置的参数具体如下。

(1)"刀具号":刀具的系列号,用于后置处理的自动换刀指令。刀具号唯一,且对应机床的刀库。

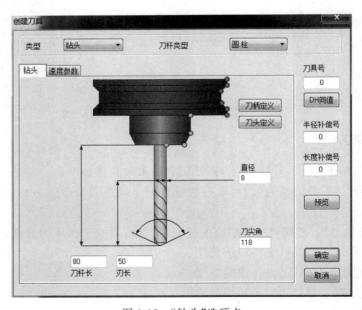

图 1-15 "钻头"选项卡

(2)"半径补偿号":刀具半径补偿值的序列号,其值对应于机床的数据库。

(3)"直径":刀具的直径。

(4)"刀尖角":钻头前端尖部的角度。

(5)"刃长":刀具的刀杆可用于切削部分的长度。

(6)"刀杆长":刀尖到刀柄之间的距离。刀杆长度应大于刀刃的有效长度。

1.5 数控车软件后置设置

后置设置是指针对不同的机床、不同的数控系统,设置特定的数控代码、数控程序格式及参数,并生成配置文件。在生成数控程序时,系统根据该配置文件的定义生成用户所需要的特定代码格式的加工指令。

后置设置给用户提供了一种灵活方便的设置系统配置的方法。针对不同的机床进行适当的配置,具有重要的实际意义。通过设置系统配置参数,后置处理所生成的数控程序可以直接输入数控机床或加工中心进行加工,而无须进行修改。如果已有的机床类型中没有所需的机床,则可增加新的机床类型以满足使用需求,并可对新增的机床进行设置。

"后置设置"对话框如图 1-16 所示,在左侧的上、下两个列表中分别列出了现有的控制系统与机床配置文件;在中间的各个选项卡中对相关参数进行设置;在右侧的"测试"栏中,可以选中刀具轨迹,并单击"生成代码"按钮,就可以在"代码"选项卡中看到当前的后置设置下,选中刀具轨迹所生成的 G 代码,便于用户对照后置设置的效果。

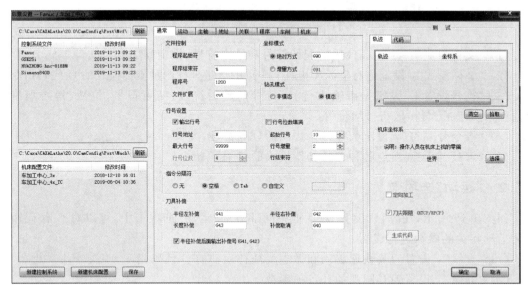

图 1-16 数控车"后置设置"对话框

操作说明:单击"数控车"标签,再单击"后置设置"按钮,弹出"后置设置"对话框,用户可按自己的需求增加新的或更改已有的控制系统和机床配置。单击"确定"按钮可将用户的更改保存;若单击"取消"按钮,则放弃已做的更改。

1.5.1 "通常"选项卡

如图 1-17 所示,在"后置设置"对话框中间部分的"通常"选项卡中,可以对 G 代码的基本格式进行设置。

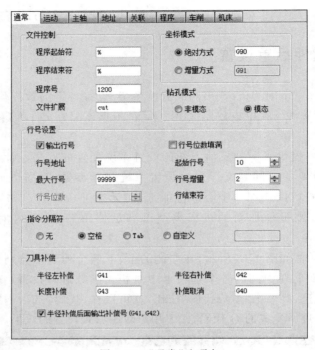

图 1-17 "通常"选项卡

(1)"文件控制":设定 G 代码的起始符号和结束符号,设定程序号及文件扩展名。
(2)"坐标模式":设定绝对坐标和相对上一点的增量坐标两种坐标模式的 G 代码指令。
(3)"行号设置":设定是否输出行号、行号的起始符号、结束符号、位数、是否填满位数、行号地址、最大行号、行号增量等。
(4)"指令分隔符":设定数控指令之间的分隔符号。
(5)"刀具补偿":设定各种刀具补偿模式的 G 代码指令。

1.5.2 "运动"选项卡

如图 1-18 所示,在"后置设置"对话框中间部分的"运动"选项卡中,可以对 G 代码中与刀具运动相关的参数进行设置。
(1)"直线":设置刀具快速移动和做直线插补运动的 G 代码指令。
(2)"圆弧":设置刀具圆弧插补各项参数。
① "代码":设置刀具做顺时针、逆时针圆弧插补运动的 G 代码指令。
② "输出平面":设置平面圆弧插补时,圆弧所在不同平面的 G 代码指令。
(3)"空间圆弧":设置空间圆弧插补的处理方式。
(4)"坐标平面圆弧的控制方式":设置在圆弧插补段的 G 代码中,圆心点(I,J,K)的坐标含义。

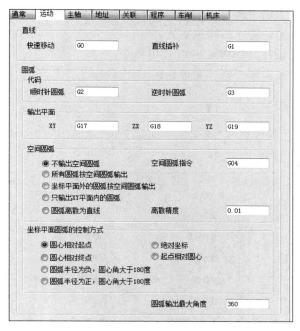

图 1-18 "运动"选项卡

1.5.3 "主轴"选项卡

如图 1-19 所示,在"后置设置"对话框中间部分的"主轴"选项卡中,可以对 G 代码中的机床主轴行为进行设置。

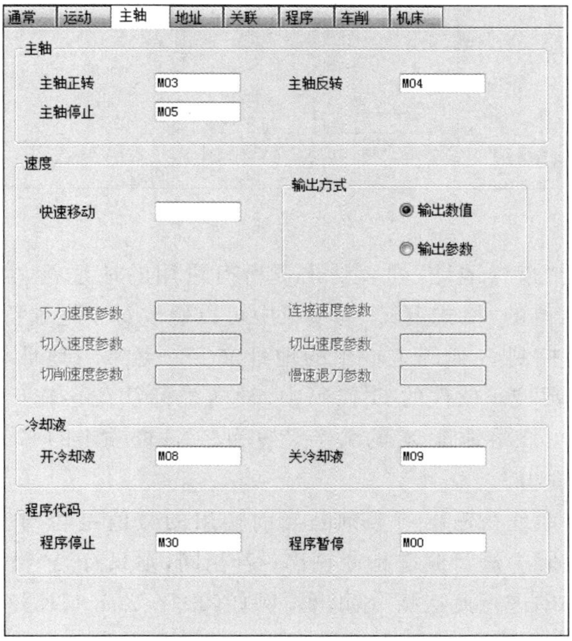

图 1-19 "主轴"选项卡

(1)"主轴":设置主轴正转、反转、停转的 G 代码指令。
(2)"速度":设置主轴转速的输出方式。
(3)"冷却液":设置开关冷却液的 G 代码指令。
(4)"程序代码":设置程序暂停和停止的 G 代码指令。

1.5.4 "地址"选项卡

如图 1-20 所示,在"后置设置"对话框中间部分的"地址"选项卡中,可以对 G 代码各指令地址的输出格式进行设置。

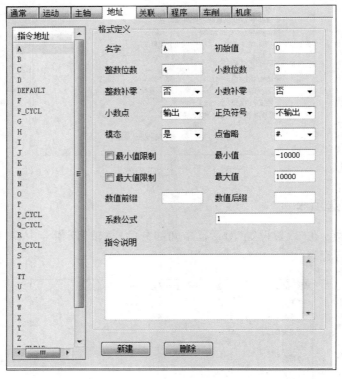

图 1-20 "地址"选项卡

(1)选项卡左侧的"指令地址"列表列出了所有可用的地址符,常用的有 X、Y、Z、I、J、K、G、M、F、S 等。在右侧的"格式定义"选项组中可以修改每个地址符的格式。

(2)"名字":直接控制 G 代码中输出的地址文字。通常与地址符自身相同,但有时需要特别设置。例如,在数控车软件的 G 代码中,轴向坐标往往会输出 Z,而在刀具轨迹中,轴向为 X 方向,因此,可以将地址 X 的名字设置为 Z,这样输出的 G 代码中,所有轨迹点的 X 坐标将用 Z 来进行输出。

(3)"模态":指令地址在输出前会判断当前输出的数值是否与上次输出的数值相同。若不同,则必须在 G 代码中进行此次指令输出;若相同,则只有当"模态"下拉列表框选择为"是"时,才会在 G 代码中进行此次指令输出。例如,X、Y、Z、I、J、K 这样的用于输出坐标的指令地址,往往在"模态"下拉列表框中选择"否",这样,若当前点 X 坐标与上一个点相同,Y 坐标不同时,此次指令在输出时将只输出新的 Y 坐标。

（4）"系数公式"：对指令地址输出的数值进行变换。例如，若将 X 指令地址的公式设置为"＊（－1）"，则所有刀位点的 X 坐标将会乘以－1 后再输出。该功能提供了一种统一修改 G 代码输出数值的可能性，但是会影响到在整个 G 代码中，所有该指令地址输出的数值，因此，在使用时，务必谨慎。

1.5.5 "关联"选项卡

如图 1-21 所示，在"后置设置"对话框中间部分的"关联"选项卡中，可以对 G 代码中各项数值在输出时，使用的指令地址进行设置。左侧的"系统变量"列表中列出了部分可以修改指令地址的数值变量。

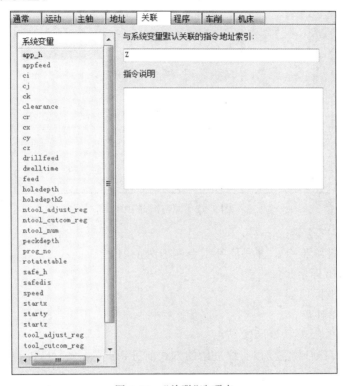

图 1-21 "关联"选项卡

1.5.6 "程序"选项卡

如图 1-22 所示，在"后置设置"对话框中间部分的"程序"选项卡中，可以对各段加工过程的 G 代码函数进行设置。

在左侧的"函数名称"列表中，列出了所有可用的函数名称，右侧的"函数体"选项卡显示了选中函数的输出格式。

例如，LatheLine 函数用于输出直线插补加工段的 G 代码，其函数体内容为"\$seq，\$speedunit，\$sgcode，\$cx，\$cz，\$feed，\$eob，@"，其中，各变量的含义如下。

（1）seq：行号。

（2）speedunit：进给速度单位。一般情况下，G98 指令代表每分钟进给量（mm/min），

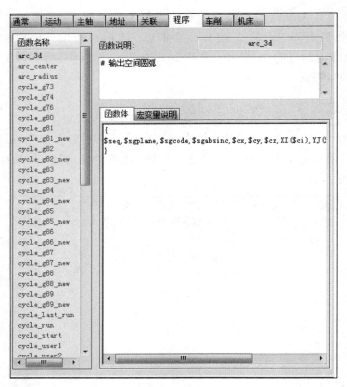

图 1-22 "程序"选项卡

G99 指令代表每转进给量(mm/r)。

(3) sgcode：进给指令。直线插补指令一般为 G01 指令。

(4) cx：径向坐标值。

(5) cz：轴向坐标值。

(6) feed：进给速度。

(7) eob：结束符,表示该函数结束。

按照以上定义,若刀具需要以直线进给的方式前进到点(50,20)处,进给速度为 20mm/min,则这段加工过程输出的 G 代码格式为

N10 G98 G01 X50.0 Z20.0 F20;

1.5.7 "车削"选项卡

如图 1-23 所示,在"后置设置"对话框中间部分的"车削"选项卡中,可以对 G 代码中车削特有的一些独特参数进行设置。

(1) "端点坐标径向分量使用直径"：轨迹中的径向坐标值使用的是半径值,但是在 G 代码中往往需要以直径值输出。在勾选此选项后,G 代码中将以直径值来输出径向坐标。例如,刀具轨迹中的径向坐标为 20,在勾选此选项后 G 代码中会输出 X40.0。

(2) "圆心坐标径向分量使用直径"：与直线插补一样,刀具轨迹中圆弧插补段的圆心坐标使用的也是半径值,若需要在 G 代码中以直径输出圆心坐标,则可以勾选此选项。在勾选此选项后,若刀具轨迹中圆心径向坐标为 20,则输出的 G 代码为 I40.0。

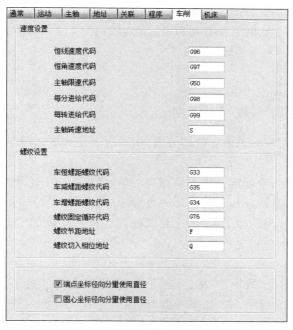

图 1-23　"车削"选项卡

1.5.8 "机床"选项卡

如图 1-24 所示,在"后置设置"对话框中间部分的"机床"选项卡中,可以对机床信息进行设置。

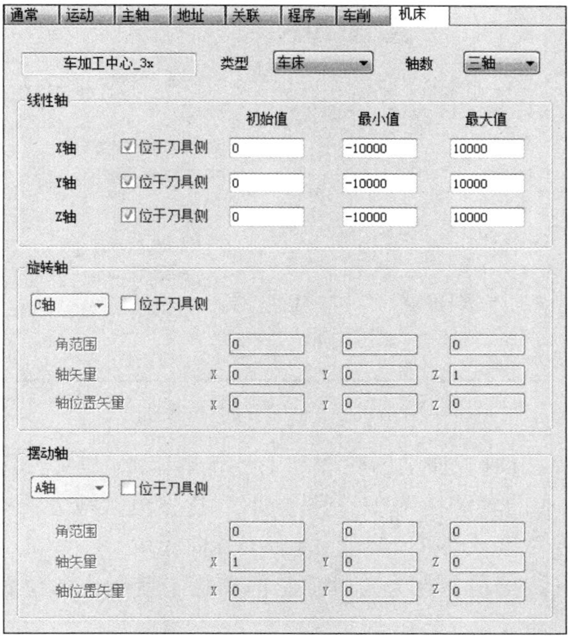

图 1-24　"机床"选项卡

如图 1-24 所示,当前选择的 3 轴车削加工中心,可以设置 3 个线形轴的初始坐标和最大、最小坐标值。若机床为 4 轴机床,则还可以设置旋转轴的相关信息,如角度范围、旋转轴向量等。

1.6 基本指令

1.6.1 "车削粗加工(创建)"对话框

车削粗加工用于实现对工件外轮廓表面、内轮廓表面和端面的粗加工,用来快速清除毛坯的多余部分。

在做轮廓粗加工时,要确定被加工轮廓和毛坯轮廓,被加工轮廓是指在加工结束后的工件表面轮廓,毛坯轮廓是指在加工前毛坯的表面轮廓。被加工轮廓和毛坯轮廓两端点相连,两轮廓共同构成一个封闭的加工区域,在此区域内的材料将被加工去除。被加工轮廓和毛坯轮廓不能单独闭合或自相交。

操作步骤如下:

(1) 在菜单栏中单击"数控车"标签,再单击"车削粗加工"按钮,弹出"车削粗加工(创建)"对话框。如图 1-25 所示,在"加工参数"选项卡中首先要确定被加工的是外轮廓表面,还是内轮廓表面或端面,接着按加工要求确定其他各加工参数。

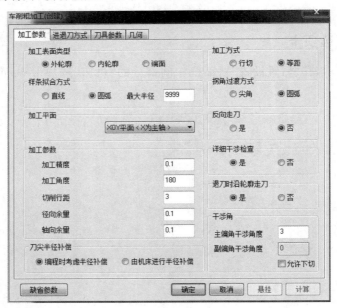

图 1-25 "加工参数"选项卡

(2) 在确定参数后,拾取被加工的轮廓和毛坯轮廓。此时,可使用系统提供的轮廓拾取工具,对于多段曲线组成的轮廓使用"限制链拾取",这将极大地方便拾取。在采用"链拾取"和"限制链拾取"时的拾取箭头方向与实际的加工方向无关。

(3) 确定进退刀点。指定一点为刀具加工前和加工后所在的位置。右击可忽略该点的输入。

在完成上述步骤后即可生成刀具轨迹。单击"数控车"标签,再单击"后置处理"按钮,拾取刚生成的刀具轨迹,即可生成加工指令。

1.6.2 "加工参数"选项卡

单击"车削粗加工(创建)"对话框中的"加工参数"标签,即进入"加工参数"选项卡。"加工参数"选项卡用于对车削粗加工中的各种工艺条件和加工方式进行限定。

各加工参数含义说明如下。

1."加工表面类型"选项组

(1)"外轮廓":采用外轮廓车刀加工外轮廓,此时默认加工方向角度为180°。

(2)"内轮廓":采用内轮廓车刀加工内轮廓,此时默认加工方向角度为180°。

(3)"端面":此时默认加工方向应垂直于系统 X 轴,即加工角度为-90°或270°。

2."加工参数"选项组

(1)"加工角度":刀具切削方向与机床 Z 轴(软件系统 X 正方向)正方向之间的夹角。

(2)"切削行距":行间切入深度,两相邻切削行之间的距离。

(3)加工余量:在加工结束后,被加工表面没有加工部分的剩余量(与最终加工结果比较)。

(4)"加工精度":用户可按需要来控制加工的精度。对轮廓中的直线和圆弧,机床可以精确地加工;对由样条曲线组成的轮廓,系统将按给定的精度把样条曲线转换成直线段来满足用户所需的加工精度。

3."拐角过渡方式"选项组

(1)"圆弧":在切削过程遇到拐角时,刀具在从轮廓的一边到另一边的过程中,以圆弧的方式过渡。

(2)"尖角":在切削过程遇到拐角时,刀具在从轮廓的一边到另一边的过程中,以尖角的方式过渡。

4."样条拟合方式"选项组

(1)"直线":对加工轮廓中的样条线,根据给定的加工精度用直线段进行拟合。

(2)"圆弧":对加工轮廓中的样条线,根据给定的加工精度用圆弧段进行拟合。

5."反向走刀"选项组

(1)"否":刀具按默认方向走刀,即刀具从机床 Z 轴正向向 Z 轴负向移动。

(2)"是":刀具按与默认方向相反的方向走刀。

6."详细干涉检查"选项组

(1)"否":假定刀具前、后干涉角度均为0°,对凹槽部分不做加工,以保证切削轨迹无前角及底切干涉。

(2)"是":在加工凹槽时,用定义的干涉角度检查加工中是否有刀具前角及底切干涉,并按定义的干涉角度生成无干涉的切削轨迹。

7."退刀时沿轮廓走刀"选项组

(1)"否":两刀位行之间首末直接进退刀,不加工行与行之间的轮廓。

(2)"是":两刀位行之间如果有一段轮廓,则在后一刀位行之前、之后增加对行间轮廓的加工。

8. "刀尖半径补偿"选项组

(1)"编程时考虑半径补偿":在生成刀具轨迹时,系统根据当前所用刀具的刀尖半径进行补偿计算(按假想刀尖点编程)。所生成的代码即为已考虑刀尖半径补偿的代码,无须机床再进行刀尖半径补偿。

(2)"由机床进行半径补偿":在生成刀具轨迹时,假设刀尖半径为0mm,按轮廓编程,不进行刀尖半径补偿计算。所生成的代码在用于实际加工时,应根据实际刀尖半径由机床指定补偿值。

9. "干涉角"选项组

(1)"主偏角干涉角度":在做前角干涉检查时,确定干涉检查的角度。

(2)"副偏角干涉角度":在做底切干涉检查时,确定干涉检查的角度。当勾选"允许下切"选项时可用。

1.6.3 "进退刀方式"选项卡

单击"车削粗加工(创建)"对话框中的"进退刀方式"标签,即进入"进退刀方式"选项卡(见图1-26)。该选项卡用于对车削粗加工中的进、退刀方式进行设定。

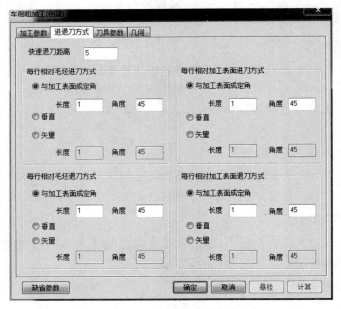

图1-26 "进退刀方式"选项卡

1. 进刀方式

"每行相对毛坯进刀方式"用于指定对毛坯部分进行切削时的进刀方式;"每行相对加工表面进刀方式"用于指定对加工表面部分进行切削时的进刀方式。

(1)"与加工表面成定角":在每一切削行前加入一段与轨迹切削方向夹角成一定角度的进刀段,刀具垂直进刀到该进刀段的起点,再沿该进刀段进刀至切削行。"角度"定义该进刀段与轨迹切削方向的夹角;"长度"定义该进刀段的长度。

(2)"垂直":指刀具直接进刀到每一切削行的起始点。

(3)"矢量":指在每一切削行前加入一段与系统 X 轴(机床 Z 轴)正方向成一定夹角的进刀段,刀具进刀到该进刀段的起点,再沿该进刀段进刀至切削行。"角度"定义矢量(进刀段)与系统 X 轴正方向的夹角;"长度"定义矢量(进刀段)的长度。

2. 退刀方式

"每行相对毛坯退刀方式"用于指定对毛坯部分进行切削时的退刀方式;"每行相对加工表面退刀方式"用于指定对加工表面部分进行切削时的退刀方式。

(1)"与加工表面成定角":指在每一切削行后加入一段与轨迹切削方向夹角成一定角度的退刀段,刀具先沿该退刀段退刀,再从该退刀段的末点开始垂直退刀。"角度"定义该退刀段与轨迹切削方向的夹角;"长度"定义该退刀段的长度。

(2)"垂直":指刀具直接进刀到每一切削行的起始点。

(3)"矢量":指在每一切削行后加入一段与系统 X 轴(机床 Z 轴)正方向成一定夹角的退刀段,刀具先沿该退刀段退刀,再从该退刀段的末点开始垂直退刀。"角度"定义矢量(退刀段)与系统 X 轴正方向的夹角;"长度"定义矢量(退刀段)的长度快速退刀距离,即以给定的退刀速度回退的距离(相对值),在此距离上以机床允许的最大进给速度进行退刀。

1.6.4 "切削用量"选项卡

在每种刀具轨迹生成时,都需要设置一些与切削用量及机床加工相关的参数。单击"刀具参数"标签并在子标签中选择"切削用量"标签可进入"切削用量"选项卡,如图 1-27 所示。

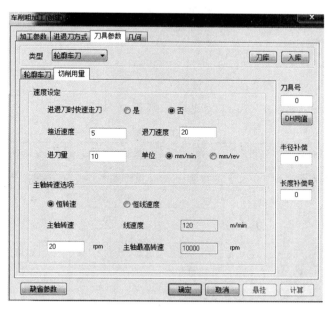

图 1-27 "切削用量"选项卡

1. "速度设定"

(1)"接近速度":刀具接近工件时的进给速度。

(2)"退刀速度":刀具离开工件的速度。

2. "主轴转速选项"

(1)"恒转速":在切削过程中,按指定的主轴转速保持主轴转速恒定,直到下一指令改变该转速。

(2)"恒线速度":在切削过程中,按指定的线速度值保持线速度恒定。

1.6.5 "轮廓车刀"选项卡

单击"刀具参数"标签可进入"轮廓车刀"选项卡。该选项卡用于对加工中所用的刀具参数进行设置。具体参数说明请参考1.4.1节中的说明。

任务二 工 艺 准 备

1.7 零件图分析

根据单向阀的使用要求,选择45钢作为本零件的毛坯材料,毛坯下料尺寸定为φ32×40。加工时以φ32毛坯外圆作为粗基准,粗加工φ20和φ30阶梯轴外圆表面及端面,然后精加工,并保证尺寸精度及表面质量,最后切断保证长度尺寸要求。

单向阀是典型的轴类零件,其包含的阶梯轴的两个外圆需要在一次装夹的情况下,同时完成两个圆柱面的加工,以保证较好的同轴度。如果分为两次定位,则零件的校正难度较大,不容易保证零件工作需要。

注意:装夹毛坯时应注意棒料伸出的长度,以免刀具与卡盘发生碰撞。

1.8 工艺设计

根据零件图分析,确定工艺过程,如表1-2所示。

表1-2 工艺过程卡片

机械加工 工艺过程卡片		产品型号	XSB	零部件序号	XSB-01	第 1 页	
		产品名称	吸水泵	零部件名称	单向阀	共 1 页	
材料牌号	C45	毛坯规格	φ32×40	毛坯质量	kg	数量	1
工序号	工序名	工序内容		工段	工艺装备	工时/min	
						准结	单件
5	备料	按φ32×40尺寸备料		外购	锯床		
10	车加工	以φ32毛坯外圆作为粗基准,粗加工φ20和φ30阶梯轴外圆表面及端面		车	车床千分尺	45	30
15	车加工	反向利用精加工过的φ20外圆和轴肩进行定位,精加工端面		车			25
20	清理	清理工件,锐角倒钝		钳			5
25	检验	检验工件尺寸		检			5

本训练任务针对第 10 和第 15 工序车削加工,进行工序设计,制订工序卡片,如表 1-3 所示。

表 1-3 车削加工工序卡片

机械加工工序卡片	产品型号	XSB-01	零部件序号	DXF-01	第 1 页
	产品名称	吸水泵	零部件名称	单向阀	共 1 页

		工序号		10、15
		工序名		车加工
		材料		C45
		设备		数控车床
		设备型号		CK6150e
		夹具		三爪卡盘
		量具		游标卡尺
				外径千分尺
		准结工时		50min
		单件工时		40min

工步	工步内容	刀具	S/(r/min)	F/(mm/r)	a_p/mm	工步工时/min	
						机动	辅助
1	工件安装						5
2	粗加工 φ20 和 φ30 外圆表面、倒角及端面,精加工余量为 0.2mm	外圆粗车刀	1200	0.2	1.5	15	
3	精加工 φ20 和 φ30 外圆表面、倒角及端面	外圆精车刀	1500	0.1	0.2	10	
4	反向装夹,粗加工 φ30 端面	外圆粗车刀	1200	0.2	1.5	10	
5	精加工 φ30 端面及倒角	外圆精车刀	1500	0.2	0.2	5	
6	拆卸、清理工件						5

技术要求:
1. 去除毛刺飞边。
2. 零件加工表面上,不应有划痕、擦伤等损伤零件表面的缺陷。
3. 未注线性尺寸公差应符合GB/T 1804—m 的要求。

任务三 软件编程训练

1.9 编程练习

1.9.1 零件造型

需对零件进行造型。此次加工为右端面和外形,建立毛坯轮廓与加工零件表面之间的封闭空间,如图1-28所示。

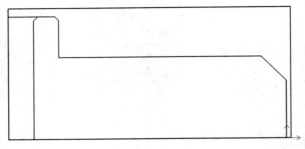

图1-28 零件造型

1.9.2 加工刀具轨迹

1. 粗加工刀具轨迹

如图1-29所示,建立轮廓车刀,根据现有刀具进行参数设置。根据车削加工工序卡片,配置参数。在选取轮廓曲线和毛坯轮廓曲线时,可在左下角命令栏中选择"单条选取"。

注意：轮廓曲线和毛坯轮廓曲线须封闭。

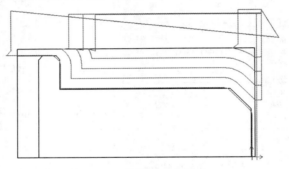

图1-29 粗加工刀具轨迹

2. 精加工刀具轨迹

如图1-30所示,所用刀具"刀尖半径补偿"均选择"由机床进行半径补偿"。这样做的好处是直接调用机床进行刀具半径补偿,方便精度调节。

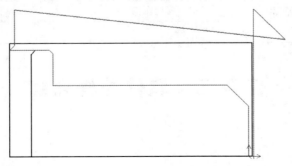

图1-30 精加工刀具轨迹

1.9.3 反向装夹造型

如图 1-31 所示,在反向装夹后,需对端面和倒角进行造型。此次加工为端面和倒角,并对毛坯轮廓进行更改。

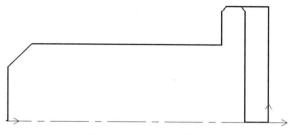

图 1-31 反向装夹造型

1.9.4 反向装夹加工刀具轨迹

1. 反向装夹粗加工刀具轨迹

反向装夹粗加工刀具轨迹如图 1-32 所示。

2. 反向装夹精加工刀具轨迹

反向装夹精加工刀具轨迹如图 1-33 所示。

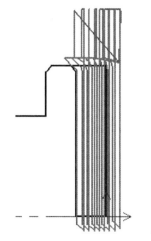

图 1-32 反向装夹粗加工刀具轨迹

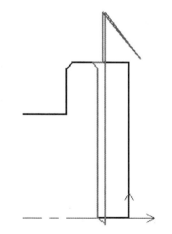

图 1-33 反向装夹精加工刀具轨迹

项 目 总 结

单向阀作为数控车床的典型阶梯轴类加工零件,被广泛应用在各种设备之中。根据设备情况和精度要求,其加工工艺也会存在一些差别。编程及操作人员需要结合加工条件合理制定加工工艺,以提高零件的加工精度和生产效率。

课后习题

1. 填空题

(1) 关于圆弧的绘制方法,数控车软件提供了_____、_____、_____、_____、_____、_____ 6种圆弧的绘制方法。

(2) 数控车基本应用界面由_____、_____、_____和_____组成。

(3) 用鼠标_____键可以激活菜单栏确定位置点或拾取元素。

(4) 数控车软件提供了_____种刀具类型设置。

(5) 如果毛坯轮廓是用来界定被加工表面的,则要求指定的轮廓是_____;如果加工的是毛坯轮廓本身,则毛坯轮廓也可以_____。

2. 判断题

(1) 在进行轮廓粗车的操作时,要确定被加工轮廓和待加工轮廓。（   ）

(2) 被加工轮廓和毛坯轮廓不能单独闭合或自相交。（   ）

(3) 恒线速度指令是在切削过程中按指定的线速度值保持线速度恒定。（   ）

(4) M09 指令是冷却液打开。（   ）

(5) 机床坐标系是指以机床原点为坐标系原点建立起来的 X 轴、Z 轴。（   ）

3. 选择题

(1) 在下列指令中,属于非模态代码的指令是（   ）。
   A. M03　　　　　B. F150　　　　　C. S250　　　　　D. G04

(2) 在数控车软件中,可以采用（   ）等方法画出圆弧。
   A. 圆心+起点　　B. 三点圆弧　　　C. 起点+半径　　D. 起点+终点

(3) 能自动捕捉直线\圆弧\圆及样条线端点的快捷键为（   ）。
   A. M 键　　　　 B. F 键　　　　　C. S 键　　　　　D. 空格键

(4) 刀库管理功能用于定义和确定刀具的有关数据,以便用户从刀库中获取刀具信息、对刀库进行维护。该功能包括（   ）的管理。
   A. 轮廓车刀　　　　　　　　　　　B. 螺纹车刀和钻头
   C. 切槽车刀　　　　　　　　　　　D. 以上都包括

(5) 钻孔加工最终所有的加工轨迹都在工件的（   ）轴上。
   A. 旋转　　　　 B. 垂直　　　　　C. 水平　　　　　D. 中心

4. 实操

(1) 数控车软件造型练习。

(2) 练习数控车粗加工加工方法。

(3) 练习数控车精加工加工方法。

（4）根据教程，尝试进行反向装夹加工编程。

自我学习检测评分表如表1-4所示。

表1-4 自我学习检测评分表

项目	目 标 要 求	分值	评分细则	得分	备注
学习关键知识点	（1）了解并熟练掌握数控车粗、精加工 （2）理解 PC 编程与手工编程的对比 （3）熟练使用数控车软件编程 （4）掌握数控车仿真刀具轨迹检查 （5）了解后置处理构造设置	50	理解与掌握		
工艺准备	（1）能够正确识读轴类零件图 （2）能够根据零件图分析，确定工艺过程 （3）能够根据工序加工工艺，正确编程并生成刀具轨迹	50	理解与掌握		

思政小课堂

项目二 泵体车削编程加工训练

➤ 思维导图

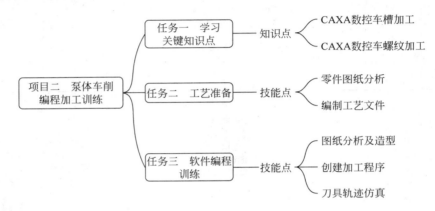

➤ 学习目标

知识目标

(1) 了解数控车槽加工。
(2) 了解数控车螺纹加工。

能力目标

(1) 能够独立确定加工工艺路线,并正确填写工艺文件。
(2) 能够正确创建刀具,并根据加工情况调整加工参数。
(3) 能够根据零件结构特点选用相应检具。

素养目标

(1) 培养学生的科学探究精神和态度。
(2) 培养学生的工程意识。
(3) 培养学生的团队合作能力。

泵体零件主要起到连接和约束自由度的作用,内孔壁需要表面粗糙度要求,尽可能减小滑动摩擦,还需保证水不从间隙流走。螺纹起连接和密封作用,需选用合适检具进行检测。由于其表面与人手有接触,所以通常要求表面光整,无毛刺,以防造成不必要的人身伤害。

项目二 泵体车削编程加工训练

▶任务引入

根据零件图(见图 2-1)要求制定加工工艺,编写数控加工程序,并完成泵体零件的加工。该零件毛坯材料为 45 钢,要求表面光整。

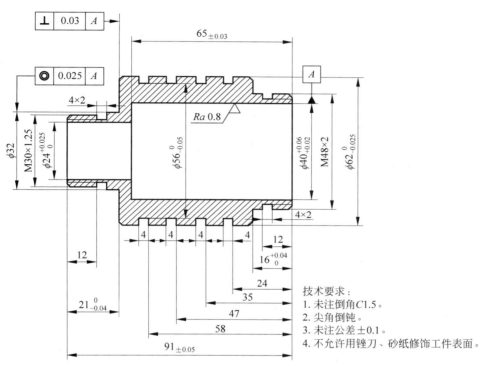

图 2-1 泵体零件图

任务一 学习关键知识点

2.1 "车削槽加工(创建)"对话框

车削槽加工用于在工件外轮廓表面、内轮廓表面和端面切槽。

切槽时要确定被加工轮廓,被加工轮廓就是加工结束后的工件表面轮廓,被加工轮廓不能闭合或自相交。

操作步骤如下。

(1) 在菜单栏中单击"数控车"标签,再单击"车削槽加工"按钮,弹出"车削槽加工(创建)"对话框。如图 2-2 所示,在"加工参数"选项卡中首先要确定被加工的是外轮廓表面,还是内轮廓表面或端面;接着按加工要求确定其他各加工参数。

(2) 在确定参数后,拾取被加工轮廓,此时,可使用系统提供的"轮廓拾取"工具。

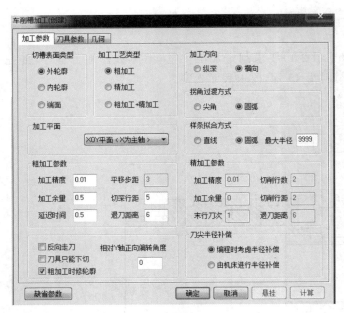

图 2-2 "加工参数"选项卡

（3）选择完轮廓后确定"进退刀点"。指定一点为刀具加工前和加工后所在的位置。右击可忽略该点的输入。

在完成上述步骤后，即可生成切槽加工轨迹。单击"数控车"标签，再单击"后置处理"按钮，拾取刚生成的切槽加工刀具轨迹，即可生成加工指令。

2.1.1 "加工参数"选项卡

加工参数主要对车削槽加工中各种工艺条件和加工方式进行限定。

各加工参数含义说明如下。

1．"切槽表面类型"选项组

（1）"外轮廓"：外轮廓切槽，或者用切槽车刀加工外轮廓。

（2）"内轮廓"：内轮廓切槽，或者用切槽车刀加工内轮廓。

（3）"端面"：端面切槽，或者用切槽车刀加工端面。

2．"加工工艺类型"选项组

（1）"粗加工"：只对槽进行粗加工。

（2）"精加工"：只对槽进行精加工。

（3）"粗加工＋精加工"：对槽进行粗加工之后，接着进行精加工。

3．"拐角过渡方式"选项组

（1）"圆弧"：当切削过程遇到拐角时，在刀具从轮廓的一边到另一边的过程中，以圆弧的方式过渡。

（2）"尖角"：当切削过程遇到拐角时，在刀具从轮廓的一边到另一边的过程中，以尖角的方式过渡。

4. "粗加工参数"选项组

（1）"延迟时间"：在粗加工槽时，刀具在槽的底部停留的时间。

（2）"切深行距"：在粗加工槽时，刀具每一次纵向切槽的切入量（机床 X 向）。

（3）"平移步距"：在粗加工槽时，刀具在切到指定的切深平移量后，进行下一次切削前的水平平移量（机床 Z 向）。

（4）"退刀距离"：在粗加工槽中，进行下一行切削前，退刀到槽外的距离。

（5）"加工余量"：在粗加工槽时，被加工表面未加工部分的预留量。

5. "精加工参数"选项组

（1）"切削行距"：在精加工槽时，行与行之间的距离。

（2）"切削行数"：在精加工槽时，刀位轨迹的加工行数，不包括最后一行的重复次数。

（3）"退刀距离"：在精加工槽中，切削完一行之后，进行下一行切削前，退刀的距离。

（4）"加工余量"：在精加工槽时，被加工表面未加工部分的预留量。

（5）"末行刀次"：在精加工槽时，为提高加工的表面质量，最后一行常常在相同进给量的情况下进行多次车削，该处定义最后一行多次切削的次数。

2.1.2 "切削用量"选项卡

车削槽加工切削用量设置如图 2-3 所示，"切削用量"选项卡的说明请参考 1.6.4 节中的说明。

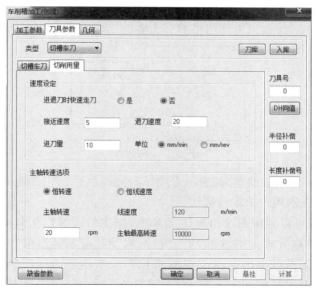

图 2-3 "切削用量"选项卡

2.1.3 "切槽车刀"选项卡

单击"刀具参数"标签可进入"切槽车刀"选项卡（见图 2-4）。该选项卡用于对加工中所用的切槽刀具参数进行设置。具体参数说明请参考 1.4.2 节中的说明。

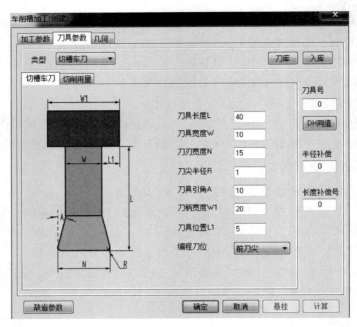

图 2-4 "切槽车刀"选项卡

2.1.4 车削槽加工实例

车削槽加工实例的步骤具体如下。

（1）如图 2-5 所示，螺纹退刀槽凹槽部分为要加工出的轮廓。

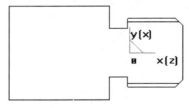

图 2-5 工件图纸

（2）填写参数表：在"切槽车刀"选项卡中填写完参数后，单击"确认"按钮。

（3）拾取轮廓，提示用户拾取轮廓线。

（4）拾取轮廓线可以利用曲线拾取工具菜单，按空格键弹出工具菜单，如图 2-6 所示。工具菜单提供三种拾取方式："单个拾取"、"链拾取"和"限制链拾取"。在拾取第一条轮廓线后，此轮廓线变色。系统给出提示："选择方向"，要求用户选择一个方向，此方向只表示拾取轮廓线的方向，与刀具的加工方向无关，如图 2-7 所示。在选择方向后，如果采用的是"链拾取"方式，则系统自动拾取首尾连接的轮廓线；如果采用"单个拾取"方式，则系统提示"继续拾取轮廓线"。此处采用"限制链拾取"方式，系统继续提示"拾取限制线"，拾取终止线段即凹槽的左边部分，凹槽部分变为虚线，如图 2-8 所示。

（5）确定"进退刀点"。指定一点为刀具加工前和加工后所在的位置。右击可忽略该点的输入。

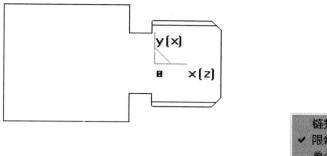

图 2-6　拾取工具菜单

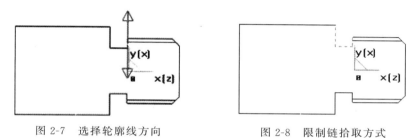

图 2-7　选择轮廓线方向　　　　图 2-8　限制链拾取方式

（6）生成刀具轨迹。在确定进退刀点之后，系统生成刀具轨迹，如图 2-9 圆框内所示。

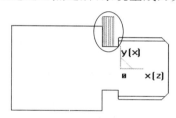

图 2-9　生成刀具轨迹

注意：被加工轮廓不能闭合或自相交。生成刀具轨迹与切槽车刀刀角半径、刀刃宽度等参数密切相关。可按实际需要只绘出退刀槽的上半部分。

2.2　"车螺纹加工（创建）"对话框

车螺纹加工为非固定循环方式加工螺纹，可对螺纹加工中的各种工艺条件、加工方式进行更为灵活的控制。

操作步骤具体如下。

（1）单击"数控车"标签，再单击"车螺纹加工"按钮，弹出"车螺纹加工（创建）"对话框。如图 2-10 所示，用户可在"螺纹参数"选项卡中确定各加工参数。

（2）拾取螺纹"起点""终点""进退刀点"。

（3）参数填写完毕，单击"确认"按钮，即生成螺纹车削刀具轨迹。

（4）单击"数控车"标签，再单击"后置处理"按钮，拾取刚生成的刀具轨迹，即可生成螺纹加工指令。

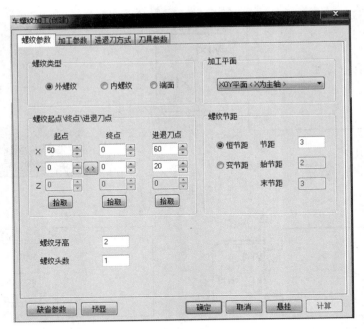

图 2-10 "螺纹参数"选项卡

2.2.1 "螺纹参数"选项卡

单击"车螺纹加工(创建)"对话框中的"螺纹参数"标签即进入"螺纹参数"选项卡。"螺纹参数"选项卡主要包括与螺纹性质相关的参数,如"螺纹类型""螺纹节距""螺纹头数"等。螺纹"起点"坐标和"终点"坐标来自前一步的拾取结果,用户也可以进行修改。

各螺纹参数含义说明如下。

(1)"起点":车螺纹加工的起点坐标,单位为 mm。

(2)"终点":车螺纹加工的终止点坐标,单位为 mm。

(3)"进退刀点":车螺纹加工进刀与退刀点的坐标,单位为 mm。

(4)"螺纹牙高":螺纹牙的高度。

(5)"螺纹头数":螺纹起始点到终止点之间的牙数。

(6)"螺纹节距":包括 5 种。"恒节距"是指两个相邻螺纹轮廓上对应点之间的距离为恒定值;"节距"是指恒定节距值;"变节距"是指两个相邻螺纹轮廓上对应点之间的距离为变化值;"始节距"是指起始端螺纹的节距;"末节距"是指终止端螺纹的节距。

2.2.2 "加工参数"选项卡

图 2-11 所示为"加工参数"选项卡,用于对螺纹加工中的工艺条件和加工方式进行设置。

各螺纹加工参数含义说明如下。

1. "加工工艺"选项组

(1)"粗加工":指直接采用粗加工方式加工螺纹。

(2)"粗加工+精加工":指根据指定的粗加工深度进行粗加工后,再采用精加工方式

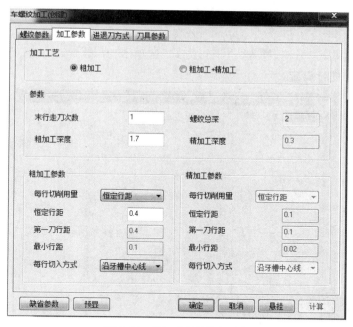

图 2-11 "加工参数"选项卡

(如采用更小的行距)切除剩余余量(精加工深度)。

2．"参数"选项组

(1)"末行走刀次数"：为提高加工表面质量，最后一行常常在相同进给量的情况下，进行多次车削，该处定义最后一行多次切削的次数。

(2)"螺纹总深"：螺纹粗加工和精加工总的切深量。

(3)"粗加工深度"：螺纹粗加工的切深量。

(4)"精加工深度"：螺纹精加工的切深量。

3．"每行切削用量"下拉列表框

(1)"恒定行距"：定义在沿恒定的行距进行加工时的行距。

(2)"恒定切削面积"：为保证每次切削的切削面积恒定，各次切削深度将逐步减小，直至等于最小行距。用户需指定第一刀行距及最小行距。吃刀深度规定如下：第 n 刀的吃刀深度为第一刀吃刀深度的 $\sqrt{n}$ 倍。

(3)"变节距"：两个相邻螺纹轮廓上对应点之间的距离为变化值。

(4)"始节距"：起始端螺纹的节距。

(5)"末节距"：终止端螺纹的节距。

4．"每行切入方式"下拉列表框

"每行切入方式"：刀具在螺纹始端切入时的切入方式。刀具在螺纹末端的退出方式与切入方式相同。

(1)"沿牙槽中心线"：在切入时沿牙槽中心线。

(2)"沿牙槽右侧"：在切入时沿牙槽右侧。

(3)"左右交替"：在切入时沿牙槽左右交替。

2.2.3 "进退刀方式"选项卡

单击"进退刀方式"标签,进入"进退刀方式"选项卡(见图 2-12)。该选项卡用于对加工中的进、退刀方式进行设定。

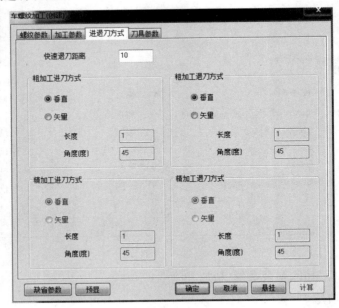

图 2-12 "进退刀方式"选项卡

1. 进刀方式

(1)"垂直":刀具直接进刀到每一切削行的起始点。

(2)"矢量":在每一切削行前加入一段与系统 X 轴(机床 Z 轴)正方向成一定夹角的进刀段,刀具进刀到该进刀段的起点,再沿该进刀段进刀至切削行。

① "长度":定义矢量(进刀段)的长度。

② "角度":定义矢量(进刀段)与系统 X 轴正方向的夹角。

2. 退刀方式

(1)垂直:刀具直接退刀到每一切削行的起始点。

(2)矢量:在每一切削行后,加入一段与系统 X 轴(机床 Z 轴)正方向成一定夹角的退刀段,刀具先沿该退刀段退刀,再从该退刀段的末点开始垂直退刀。

① 长度:定义矢量(退刀段)的长度。

② 角度:定义矢量(退刀段)与系统 X 轴正方向的夹角。

3. "快速退刀距离"

"快速退刀距离":以给定的退刀速度回退的距离(相对值),在此距离上以机床允许的最大进给速度退刀。

2.2.4 "切削用量"选项卡

单击"切削用量"标签可进入"切削用量"选项卡(见图 2-13)。"切削用量"选项卡的说

项目二 泵体车削编程加工训练

明参考1.6.4节中的说明。

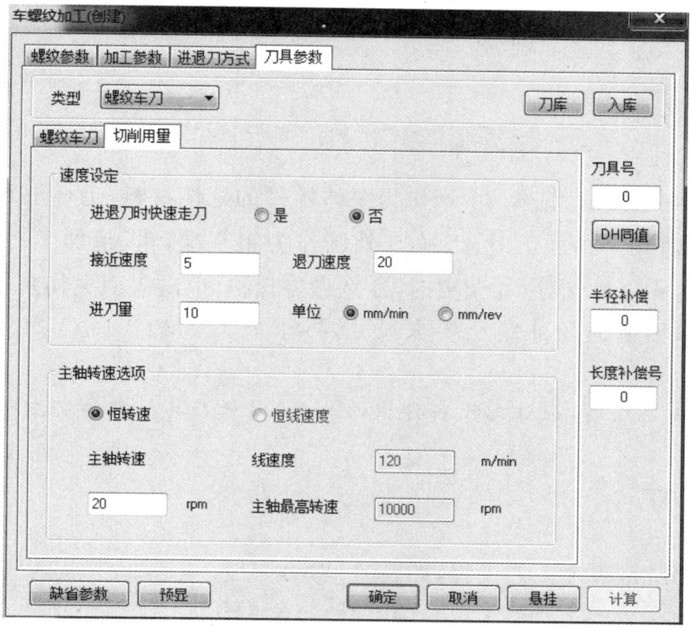

图2-13 "切削用量"选项卡

2.2.5 "螺纹车刀"选项卡

单击"螺纹车刀"标签可进入"螺纹车刀"选项卡(见图2-14)。该选项卡用于对加工中所用的螺纹车刀参数进行设置。具体参数说明请参考1.4.3节中的说明。

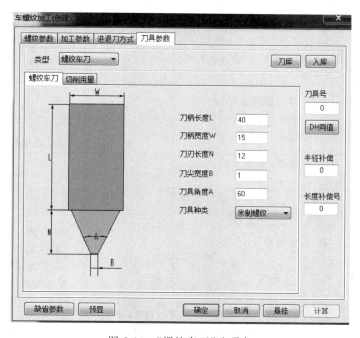

图2-14 "螺纹车刀"选项卡

任务二 工 艺 准 备

2.3 零件图分析

根据零件的使用要求，选择 45 钢作为泵体零件的毛坯材料，毛坯下料尺寸定为 $\phi 65 \times 100$，中间有 $\phi 20$ 通孔。在加工时，以 $\phi 65$ 外圆作为粗基准，粗、精加工右侧部分至要求尺寸，然后掉头装夹 $\phi 62$ 外圆处（在装夹时，注意做好保护，以防表面夹伤），加工零件左端 $\phi 32$ 和 M30 螺纹阶梯轴端面及外圆至要求尺寸，再车加工 M30×1.25 螺纹，用螺纹通止规检验。

注意：在装夹毛坯时，应注意棒料伸出的长度，以免刀具与卡盘发生碰撞。

2.4 工艺设计

根据零件图分析，确定工艺过程，如表 2-1 所示。

表 2-1 工艺过程卡片

机械加工 工艺过程卡片	产品型号	XSB	零部件序号	XSB-02	第 1 页	
	产品名称	吸水泵	零部件名称	泵体	共 1 页	
材料牌号	C45	毛坯规格	$\phi 65 \times 100$	毛坯质量	kg	数量	1

工序号	工序名	工序内容	工段	工艺装备	工时/min	
					准结	单件
5	备料	按 $\phi 65 \times 100$ 尺寸备料，中间有 $\phi 20$ 通孔	外购	锯床		
10	车加工	以 $\phi 65$ 毛坯外圆作为粗基准，精加工 $\phi 62$ 和 M48 螺纹大径外圆及端面	车	车床 千分尺	130	30
15	车加工	切螺纹退刀槽和 $\phi 62$ 圆上槽		车床 游标卡尺		10
20	车加工	加工螺纹		车床 螺纹通止规		10
25	车加工	打内孔底孔，精镗 $\phi 40$ 和 $\phi 24$ 内孔		车床 内径千分尺		20
30	车加工	反向装夹，以 $\phi 62$ 外圆为精基准，加工螺纹大径和端面		车床 千分尺		20
35	车加工	切螺纹退刀槽		车床 游标卡尺		10
40	车加工	加工 M30 螺纹		车床 螺纹通止规		10
45	清理	清理工件，锐角倒钝	钳			5
46	检验	检验工件尺寸	检			5

本训练任务针对第10～第40工序车削加工,进行工序设计,制订工序卡片,如表2-2所示。

表 2-2 车削加工工序卡片

机械加工工序卡片	产品型号	XSB	零部件序号	XSB-02	第 1 页
	产品名称	吸水泵	零部件名称	泵体	共 1 页

工序号		10～40
工序名		车加工
材料		C45
设备		数控车床
设备型号		CK6150e
夹具		三爪卡盘
量具		游标卡尺、千分尺 内径千分尺 螺纹通止规
准结工时		125min
单件工时		105min

技术要求:
1. 未注倒角C1.5。
2. 尖角倒钝。
3. 未注公差±0.1。
4. 不允许用锉刀、砂纸修饰工件表面。

工步	工步内容	刀具	S/(r/min)	F/(mm/r)	a_p/mm	工步工时/min 机动	工步工时/min 辅助
1	工件安装						5
2	粗加工 $\phi 62$ 和 $\phi M48$ 外圆表面、倒角及端面,精加工余量为 0.2mm	外圆粗车刀	1200	0.2	1.5	15	
3	精加工 M48 螺纹大径,精加工 $\phi 62$ 外圆、端面	外圆精车刀	1500	0.1	0.2	10	
4	切槽加工,切螺纹退刀槽,切 $\phi 62$ 外圆上槽	切槽刀	600	0.1		15	
5	粗车螺纹	螺纹车刀	600	2	0.4	5	
6	精车螺纹	螺纹车刀	600	2	0.1	5	

续表

工步	工步内容	刀具	S/(r/min)	F/(mm/r)	a_p/mm	工步工时/min 机动	工步工时/min 辅助
7	粗车内孔	内孔粗车刀	1200	0.2	1.5	10	
8	精车内孔	内控精车刀	1500	0.1	0.2	5	
9	反向装夹、找正						10
10	粗加工 M30 螺纹大径及端面和倒角	外圆粗车刀	1200	0.2	1.5	15	
11	精加工 M30 螺纹大径及端面和倒角	外圆精车刀	1500	0.1	0.2	10	
12	切螺纹退刀槽	切槽刀	600	0.1		5	
13	粗车螺纹、精车螺纹	螺纹刀	600	1.5	0.4	10	
14	拆卸、清理工件						5

任务三　软件编程训练

2.5　编程练习

2.5.1　零件造型

如图 2-15 所示，画出毛坯轮廓，建立毛坯轮廓与加工零件表面之间的封闭空间，螺纹处将倒角补齐，方便进刀和定位螺纹起点。

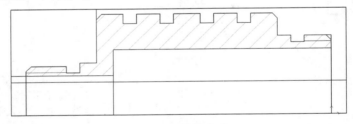

图 2-15　零件造型

2.5.2　加工刀具轨迹

如图 2-16～图 2-19 所示，逐步建立轮廓车刀，根据现有刀具进行参数设置，根据车削工艺卡配置参数。

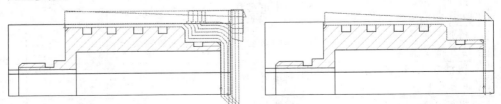

图 2-16　粗、精加工外圆轮廓刀具轨迹

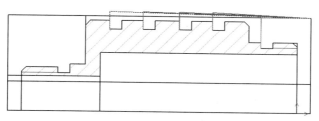

图 2-17 切槽路径刀具轨迹

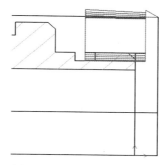

图 2-18 螺纹车削刀具轨迹

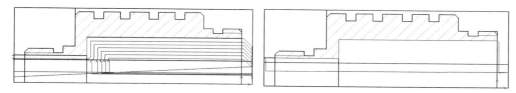

图 2-19 内孔车削粗、精加工刀具轨迹

2.5.3 反向装夹造型

如图 2-20 所示，在反向装夹后，需对端面和倒角进行造型。此次加工为端面和倒角，并对毛坯轮廓进行更改。

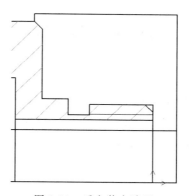

图 2-20 反向装夹造型

2.5.4 反向装夹加工刀具轨迹

反向装夹加工刀具轨迹如图 2-21～图 2-23 所示。

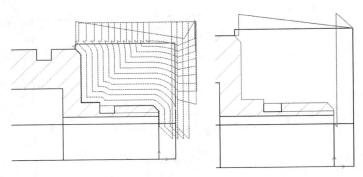

图 2-21　反向装夹粗、精加工 M30 螺纹大径及端面刀具轨迹

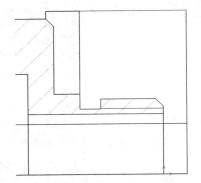

图 2-22　反向装夹加工螺纹退刀槽刀具轨迹

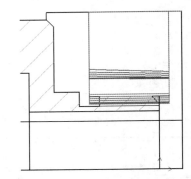

图 2-23　反向装夹螺纹车加工刀具轨迹

项 目 总 结

泵体作为数控车床的腔体典型加工零件，在生产和生活中应用广泛。根据设备情况和精度的要求，其加工工艺也存在一些差别。编程人员及操作人员需要结合加工条件，合理制定加工工艺，以提高零件的加工精度和生产效率。

课 后 习 题

1. 填空题

（1）在 CAXA 电子图板中，保存的图形文件的后缀是_____。

（2）正交线表示所绘制直线为与坐标轴_____的线段，要么水平、要么竖直。

（3）车削槽加工在数控车软件设置上可分为 3 个步骤，分别是_____、_____、_____。

（4）常用于粗、精加工的加工指令是_____、_____、_____、_____；其中，粗加工指令是_____；精加工指令是_____。

（5）螺纹参数中螺纹节距包含_____、_____、_____、_____。

2. 判断题

（1）在数控车中"链拾取"无须用户指定起始曲线及链搜索方向。　　　　　　　　（　　）

(2) 由于数控机床的先进性,因此,任何零件均适合在数控机床上加工。（　　）
(3) 数控车一般规定 G86 指令为螺纹车削循环指令。（　　）
(4) 在固定循环功能指令中的 K 代码是指重复加工次数,一般在增量方式下使用。
（　　）
(5) 为保证每次切削的切削面积恒定,各次切削深度将逐步减小,直至等于最小行距。
（　　）

3. 选择题

(1) 拾取轮廓线可以利用曲线拾取工具菜单,按空格键弹出工具菜单,工具菜单提供三种拾取方式,以下错误的是(　　)。
　　A. 单个拾取　　　B. 链拾取　　　C. 限制链拾取　　　D. 随意拾取
(2) 在拾取第一条轮廓线后,此轮廓线变为(　　)。
　　A. 红色虚线　　　B. 红色实线　　　C. 蓝色虚线　　　D. 蓝色实线
(3) 下列生成刀具轨迹的注意事项中,错误的是(　　)。
　　A. 被加工轮廓不能闭合或自相交
　　B. 生成轨迹与切槽车刀刀角半径、刀刃宽度等参数密切无关
　　C. 可按实际需要只绘出退刀槽的上半部分
　　D. 以上均错误
(4) 在车螺纹加工中,不需要确定的点是(　　)。
　　A. 螺纹起点　　　B. 螺纹终点　　　C. 螺纹中点　　　D. 进退刀点
(5) 在车螺纹加工中各次切削深度将逐步减小。吃刀深度规定如下:第 n 刀的吃刀深度为第一刀吃刀深度的(　　)倍。
　　A. $n/2$　　　B. $\sqrt{n}$　　　C. $\sqrt[3]{n}$　　　D. $\sqrt[4]{n}$

4. 简答题

(1) 简述如何调节计算机编程加工精度。

(2) 简述车螺纹加工左旋螺纹的方法及步骤。

(3) 车削槽加工还能加工哪些类型的零件？

自我学习检测评分表如表2-3所示。

表2-3 自我学习检测评分表

项目	目标要求	分值	评分细则	得分	备注
学习关键知识点	(1) 掌握车削槽加工技能 (2) 掌握刀尖圆弧半径补偿指令的使用 (3) 理解螺纹加工参数设置	20	理解与掌握		
工艺准备	(1) 能够正确识读零件图 (2) 能够独立确定加工工艺路线,并正确填写工艺文件 (3) 能够根据工序加工工艺,编写正确的加工程序	30	理解与掌握		
编程练习	(1) 编程步骤完整无缺失 (2) 生成刀具轨迹无错误 (3) 参数匹配合理	50	掌握		

思政小课堂

项目三　曲柄车削编程加工训练

➢ 思维导图

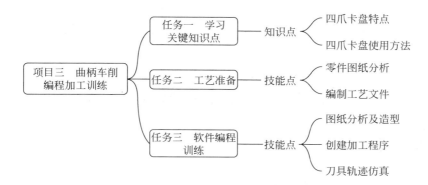

➢ 学习目标

知识目标

(1) 了解轴类零件加工精度,合理安排加工工艺。
(2) 将理论联系实际,灵活运用知识解决问题。

能力目标

(1) 能够独立确定加工工艺路线,并正确填写工艺文件。
(2) 能够正确判断工件工序是否合理。
(3) 能够根据零件结构特点和精度,合理选用合适的加工方法和加工方案。

素养目标

(1) 培养学生的科学探究精神和态度。
(2) 培养学生的工程意识。
(3) 培养学生的团队合作能力。

曲柄类零件广泛地应用于生产和生活中,主要由阶梯轴部分和偏心轴部分构成。偏心轴部分用于做变心运动。

➢ 任务引入

根据曲柄零件图(见图3-1)要求,制定加工工艺、编写数控加工程序,并完成曲柄零件的加工。该零件毛坯材料为45钢,要求表面光整。

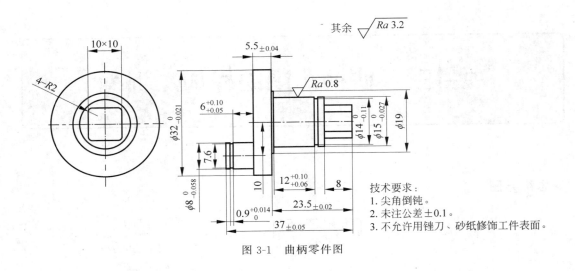

图 3-1 曲柄零件图

任务一　学习关键知识点

3.1　四爪卡盘简介

3.1.1　介绍

四爪单动卡盘的全称是机床用手动四爪单动卡盘,是由 1 个盘体、4 个丝杆、1 副卡爪组成的。在工作时是用 4 个丝杠分别带动四爪,因此,常见的四爪单动卡盘没有自动定心的功能。

随着人们对工作效率的要求越来越高,液压卡盘逐渐取代手动卡盘,得到了广泛应用。为满足不同工件的加工要求,如矩形面、圆柱毛坯面等一些不规则形面的工件,以及一些夹持面与加工面存在偏心的工件,通常需要采用四爪液压卡盘。纵观国内、外液压卡盘生产厂家生产的四爪液压卡盘,其结构均为一个油缸驱动 4 个卡爪同时夹紧,工件中心很难与主轴回转中心同轴,每次夹持中心的调整都需通过卡爪来实现,很不方便。即使调整得比较精确,但由于工件自身夹持面的偏差,根据三点定心的原理,仍旧不能实现四点同圆,因此,在 4 个卡爪中三爪夹紧后,第 4 爪为虚夹。当切削力较大时,夹持力可能不够,造成工件报废甚至发生事故,这些制约了四爪液压卡盘的应用。

四爪自定心卡盘的全称是机床用手动四爪自定心卡盘,是由 1 个盘体、4 个小伞齿、1 副卡爪组成的。4 个小伞齿和盘丝啮合,盘丝的背面有平面螺纹结构,卡爪等分安装在平面螺纹上。当用扳手扳动小伞齿时,盘丝便转动,它背面的平面螺纹就使卡爪同时向中心靠近或退出。因为盘丝上的平面矩形螺纹的螺距相等,所以四爪运动距离相等,有自动定心的作用。

四爪自定心卡盘的卡爪有两种:整体爪与分离爪。整体爪是基爪和顶爪为一体的卡爪,1 副整体爪分为 4 个正爪,4 个反爪;而 1 副分离爪只有 4 个卡爪,每个卡爪都是由基爪与顶爪构成的,通过顶爪的变换,达到正爪和反爪的功用。此外,还可根据用户要求提供软卡爪,经随机配车(磨)后可获得较高定心精度,满足夹持要求。

四爪单动卡盘的卡爪只有一种整体爪。一副卡爪可单独移动,适用于夹持偏心零件和不规则形状零件。

3.1.2 适用范围

1. 四爪自定心卡盘

功能:四爪同步移动适用于夹持四方形零件,也适用于轴类、盘类零件。

适用机床及附件:普通车床、经济型数控车床、磨床、铣床、钻床及机床附件——分度头回转台等。

2. 四爪单动卡盘

功能:每一个卡爪都可单独移动适用于夹持偏心零件和不规则形状零件。

适用机床及附件:普通车床、经济型数控车床、磨床、铣床、钻床及机床附件——分度头回转台等。

3.1.3 用途

车床附件,用以夹持圆形或方形、矩形工件,进行切削加工。四爪卡盘的四爪不能联动,需分别扳动,故还能用来夹持单边、不对中心的零件。

3.2 车削偏心零件的加工方法

3.2.1 利用三爪卡盘装夹

1. 车削方法

对于图 3-2(a)所示的长度较短的偏心工件,可以在三爪卡盘上进行车削。先把偏心工件中的非偏心部分的外圆车好,随后在卡盘任意一个卡爪与工件接触面之间,垫上一块预先选好厚度的垫片,经校正母线与偏心距,并把工件夹紧后,即可车削。

在一般情况下,垫片厚度可用近似公式计算

$$x = 1.5e(偏心距)$$

若想计算得更精确一些,则需在近似公式中代入偏心距修正值 k 来计算和调整垫片厚度,则近似公式为

$$x = 1.5e + k$$
$$k \approx 1.5\Delta e$$
$$\Delta e = e - e_{测}$$

式中:e——工件偏心距;

k——偏心距修正值,其正负按实测结果确定;

Δe——试切后实测偏心距误差;

$e_{测}$——在试切后,实测偏心距。

2. 偏心工件的测量、检查

在工件调整校正母线和偏心距时,主要是采用带有磁力表座的百分表在车床上进行校

正,如图 3-2(b)所示,直至符合要求后方可进行车削。待工件加工完成后,为确定偏心距是否符合要求,还需进行最后检查。检查方法是把工件放入 V 形铁中,用百分表在偏心圆处测量,缓慢转动工件,观察其跳动量。

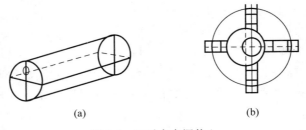

图 3-2　三爪卡盘调偏心
(a) 偏心工件外形；(b) 车床校正偏心

3.2.2　利用四爪单动卡盘装夹

找正步骤具体如下。

(1) 把划好线的工件装在四爪单动卡盘上。在装夹时,先调节卡盘的两爪,使其呈不对称位置,另两爪呈对称位置,工件偏心圆线在卡盘中央,如图 3-3(a)所示。

(2) 在床面上放好小平板和划针盘,针尖对准偏心圆线,校正偏心圆。然后把针尖对准外圆水平线,如图 3-3(b)所示,自左至右检查水平线是否水平。把工件转动 90°,用同样的方法检查另一条水平线,然后紧固卡脚和复查工件装夹情况。

(3) 在工件校准后,把四爪再拧紧一遍,即可进行车削。在初车削时,进给量要小,车削深度要浅,等工件车圆后切削用量可以适当增加,否则会损坏车刀或使工件移位,如图 3-3(a)所示。

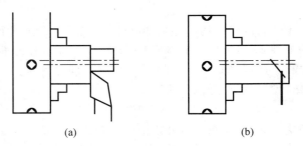

图 3-3　四爪单动卡盘调偏心

上述两种方法都是比较常用的加工方法,但是都存在缺点：装夹比较麻烦、不容易校正、容易产生误差,而且不适合批量生产。针对上述缺点,设计了一种专门用于批量生产的偏心夹具,并已投入生产,达到了一定的效果。

3.2.3　偏心轮车夹具

偏心轮车夹具主要用于批量生产偏心零件,图 3-4 所示为该夹具零件图。

在装夹时打表校正 $\phi60$ 的内孔,保证零件形状、位置、公差的要求。装夹方法为夹具的左端装夹工件,通过百分表校正,右端装夹在三爪自定心卡盘上。

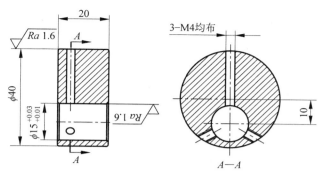

图 3-4 偏心轮车夹具零件图

任务二 工 艺 准 备

3.3 零件图分析

根据零件的使用要求,选择 45 钢作为曲柄零件的毛坯材料,毛坯下料尺寸定为 $\phi 35 \times 40$。在加工时,以 $\phi 35$ 毛坯外圆作为粗基准,粗、精加工右侧端面、$\phi 14$ 和 $\phi 15$ 圆柱表面至要求尺寸,切槽,然后掉头,四爪装夹 $\phi 15$ 外圆处(装夹时注意做好保护,以防表面夹伤),加工零件左端 $\phi 8$ 偏心轴至要求尺寸,切槽。

注意:在车削右侧 $\phi 32$ 外圆时,车削长度要足够。另外,在装夹毛坯时,应注意棒料伸出的长度,以免刀具与卡盘发生碰撞。

3.4 工艺设计

根据零件图分析,确定工艺过程,如表 3-1 所示。

表 3-1 工艺过程卡片

机械加工 工艺过程卡片		产品型号	XSB	零部件序号	XSB-03	第 1 页	
		产品名称	吸水泵	零部件名称	曲柄	共 1 页	
材料牌号	C45	毛坯规格	$\phi 35 \times 40$	毛坯质量	kg	数量	1
工序号	工序名	工序内容		工段	工艺装备	工时/min	
						准结	单件
5	备料	按 $\phi 35 \times 40$ 尺寸备料		外购	锯床		
10	车加工	以 $\phi 35$ 毛坯外圆作为粗基准,精加工 $\phi 15$ 和 $\phi 32$ 外圆及端面		车	车床 千分尺	100	20
15	车加工	切卡簧槽			车床 游标卡尺		10
20	车加工	以 $\phi 32$ 外圆为精基准反向装夹,调偏心距			车床		20

续表

工序号	工序名	工序内容	工段	工艺装备	工时/min 准结	工时/min 单件
25	车加工	精加工φ8外圆及端面		车床 千分尺		20
30	车加工	切卡簧槽		车床 游标卡尺		10
35	清理	清理工件,锐角倒钝	钳			5
40	检验	检验工件尺寸	检			5

本训练任务针对第10~第30工序车削加工,进行工序设计,制订工序卡片,如表3-2所示。

表3-2 车削加工工序卡片

机械加工工序卡片		产品型号	XSB	产品型号	XSB-03	第1页
		产品名称	吸水泵	产品名称	曲柄	共1页

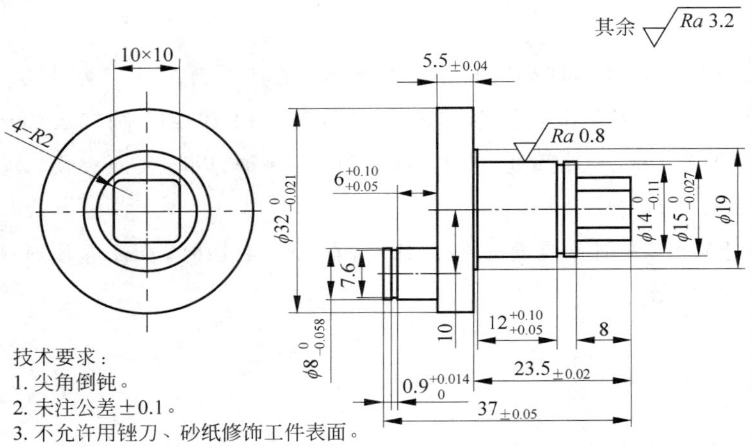

工序号	10~30
工序名	车加工
材料	C45
设备	数控车床
设备型号	CK6150e
夹具	四爪卡盘
量具	千分尺 游标卡尺 千分尺
准结工时	90min
单件工时	80min

技术要求:
1. 尖角倒钝。
2. 未注公差±0.1。
3. 不允许用锉刀、砂纸修饰工件表面。

工步	工步内容	刀具	S/(r/min)	F/(mm/r)	a_p/mm	工步工时/min 机动	工步工时/min 辅助
1	工件安装						5
2	粗加工φ32、φ15外圆表面、倒角及端面,精加工余量为0.2mm	外圆粗车刀	1200	0.2	1.5	15	
3	精加工φ32、φ15外圆表面、倒角及端面	外圆精车刀	1500	0.1	0.2	10	
4	切槽加工,切卡簧槽	切槽刀	600	0.1		5	
5	反向装夹,调偏心距						20
6	粗加工φ8外圆及端面	外圆粗车刀	1200	0.2	1.5	15	
7	精加工φ8外圆及端面	外圆精车刀	1500	0.2	0.2	5	
8	拆卸、清理工件						5

任务三　软件编程训练

3.5　编程练习

3.5.1　零件造型

零件造型如图 3-5 所示。

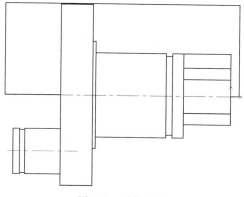

图 3-5　零件造型

3.5.2　加工刀具轨迹

加工刀具轨迹如图 3-6 和图 3-7 所示。

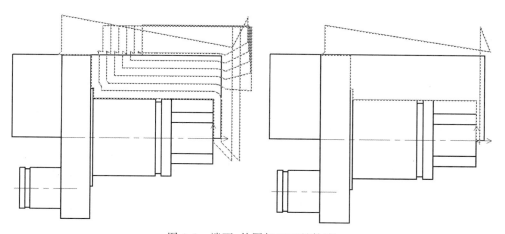

图 3-6　端面、外圆加工刀具轨迹

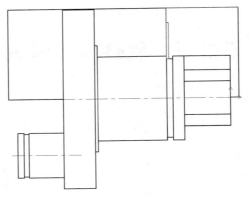

图 3-7　卡簧槽切槽刀具轨迹

3.5.3　反向装夹调偏心后造型及加工刀具轨迹

反向装夹调偏心后造型如图 3-8 所示,其外圆和端面加工刀具轨迹如图 3-9 所示,卡簧槽切槽刀具轨迹如图 3-10 所示。

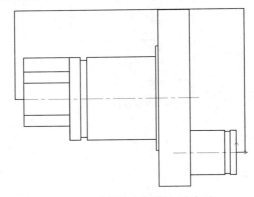

图 3-8　反向装夹调偏心后造型

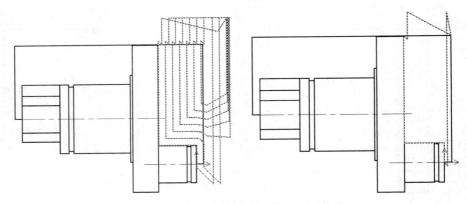

图 3-9　反向装夹外圆、端面加工刀具轨迹

注意：此曲柄零件存在外四方形状,目前车床不能加工,须上加工中心进行铣削加工,故在车床加工时留有加工余量。

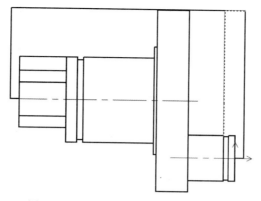

图 3-10　反向装夹卡簧槽切槽刀具轨迹

项 目 总 结

曲柄作为数控车床的典型加工零件,在生产和生活中应用广泛。根据设备情况和精度的要求,其加工工艺也存在一些差别。编程人员及操作人员需要结合加工条件,合理制定加工工艺,以提高零件的加工精度和生产效率。

课 后 习 题

1. 填空题

（1）四爪单动卡盘全称是机床用手动四爪单动卡盘,是由1个_____、4个_____、1副_____组成的。

（2）四爪自定心卡盘的卡爪分为两种:_____与_____。

（3）四爪单动卡盘适用于夹持_____和_____零件。

（4）加工偏心圆中垫片厚度的近似公式为:_____。

（5）偏心轮车夹具主要是用于_____。

2. 判断题

（1）四爪液压卡盘应用中当切削力较大时,夹持力可能不够,造成废品甚至发生事故,制约四爪液压卡盘的应用。　　　　　　　　　　　　　　　　　　　　　　（　）

（2）四爪自定心卡盘适用于夹持四方、四方形零件,也适用于轴类、盘类零件。（　）

（3）偏心圆加工垫片厚度的近似公式中 e 表示偏心距修正值。　　　　　　（　）

（4）偏心工件的检测方法是把工件放入V形铁中,用百分表在偏心圆处测量,缓慢转动工件,观察其跳动量。　　　　　　　　　　　　　　　　　　　　　　　（　）

（5）鉴于常用四爪夹具在加工偏心圆时装夹麻烦,容易产生误差,偏心夹具很好地解决了这些问题,且适用于批量生产。　　　　　　　　　　　　　　　　　　　（　）

3. 选择题

（1）车床上的三爪卡盘和铣床上的平口钳属于(　　　)。

A. 通用夹具　　　　B. 专用夹具　　　　C. 组合夹具　　　　D. 随行夹具

(2) 组合夹具不适用于(　　)生产。

A. 单件小批量　　　　　　　　　　B. 位置精度高的工件

C. 大批量　　　　　　　　　　　　D. 新产品试制

(3) 四爪单动卡盘与四爪自定心卡盘在构成上的区别是(　　)。

A. 盘体　　　　B. 夹紧装置　　　　C. 活动卡爪　　　　D. 驱动装置

(4) 在偏心工件的测量中,利用四爪单动卡盘装夹找正步骤共(　　)部分。

A. 1　　　　　　B. 2　　　　　　C. 3　　　　　　D. 4

(5) 以下不属于偏心轮夹具的优点是(　　)。

A. 装夹比较麻烦　　　　　　　　　B. 容易找正

C. 精度较高　　　　　　　　　　　D. 适用于批量生产

4. 简答题

(1) 简述偏心零件的车削加工方法。

(2) 简述如何加工 11×11 的四方形。

(3) 实操学习四爪卡盘调偏心距。

(4) 根据图 3-11 所示的曲柄零件图的要求,预习制造工程师软件的内容。

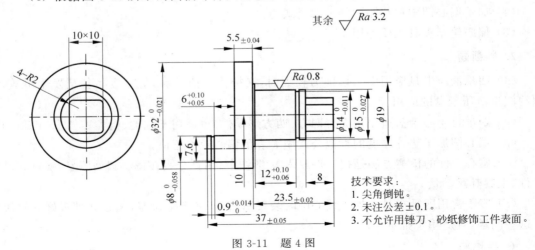

图 3-11　题 4 图

自我学习检测评分表如表 3-3 所示。

表 3-3　自我学习检测评分表

项目	目标要求	分值	评分细则	得分	备注
学习关键知识点	（1）了解偏心轴车削工艺 （2）掌握偏心轴车削加工方法 （3）学会用基础知识解决复杂轴类零件加工的问题	20	理解与掌握		
工艺准备	（1）能够正确识读零件图 （2）能够独立确定加工工艺路线，并正确填写工艺文件 （3）能够根据工序加工工艺，编写正确的加工程序	30	理解与掌握		
编程训练	（1）能够完成复杂零件的加工编程 （2）熟练使用加工坐标系变换指令 （3）合理配置机床加工参数	50	（1）理解与掌握 （2）操作流程		

思政小课堂

项目四　支撑座铣削编程加工训练

➢ **思维导图**

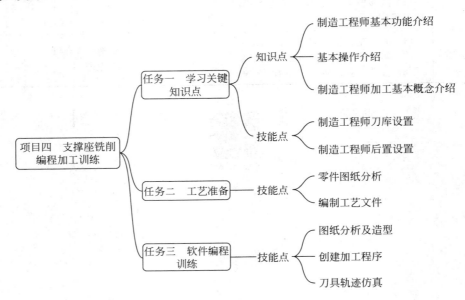

➢ **学习目标**

知识目标

(1) 了解制造工程师 2020 软件操作界面。
(2) 掌握制造工程师 2020 软件基本绘图和造型功能。
(3) 掌握平面区域粗加工、平面轮廓精加工功能。

能力目标

(1) 掌握基本准备指令和辅助指令的使用方法。
(2) 掌握铣削刀具的选择与参数设置。
(3) 能够独立确定加工工艺路线,并正确填写工艺文件。
(4) 能够根据零件结构特点和精度,合理选择加工方法和精度控制。

素养目标

(1) 培养学生的科学探究精神和态度。
(2) 培养学生的工程意识。
(3) 培养学生的团队合作能力。

任务一 学习关键知识点

4.1 制造工程师软件基本功能介绍

制造工程师2020是基于CAXA 3D实体设计2020平台,全新开发的CAD/CAM一体化系统。在建模方面,采用了精确的特征实体造型技术,同时,继承和发展了制造工程师软件以前版本的线架、曲面造型功能;在加工方面,涵盖了从两轴到五轴的数控铣功能,将三维CAD模型与CAM加工技术无缝集成;支持先进实用的轨迹参数化和批处理功能,支持高速切削,提供了知识加工功能、通用后置处理,还包含大量设计元素库。

4.2 基本操作介绍

制造工程师2020的工作窗口如图4-1所示(本书在建模方面默认已掌握设计基础)。

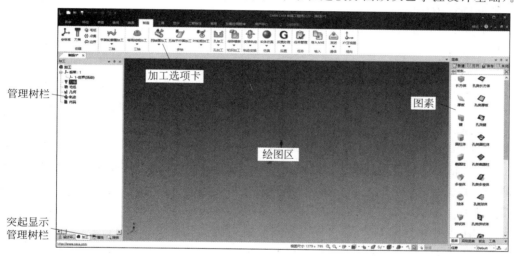

图4-1 制造工程师2020工作窗口

4.3 制造工程师加工基本概念介绍

4.3.1 造型

实体造型主要有拉伸、旋转、导动、放样、倒角、圆角、打孔、筋板、拔模、分模等特征造型方式。可以将二维的草图轮廓快速生成三维实体模型。提供多种构建基准面的功能,用户可以根据已知条件构建各种基准面。

曲面造型提供多种非均匀有理B样条(non-uniform rational B-splines,NURBS)曲面造型手段:可通过扫描、放样、旋转、导动、等距、边界和网格等多种形式生成复杂曲面;并提供曲面线裁剪和面裁剪、曲面延伸、按照平均切矢或选定曲面切矢的曲面缝合功能、多张曲面之间的拼接功能；另外，提供强大的曲面过渡功能，可以实现两面、三面、系列面等曲面过渡方式，还可以实现等半径或变半径过渡。

系统支持实体与复杂曲面混合的造型方法，可用于复杂零件设计或模具设计。系统还提供曲面裁剪实体功能、曲面加厚成实体、闭合曲面填充生成实体功能。另外，系统还允许将实体的表面生成曲面供用户直接引用。

曲面造型和实体造型方法的完美结合，是制造工程师软件在 CAD 上的一个突出特点。每一个操作步骤，软件的提示区都有操作提示功能，不管是初学者还是具有丰富 CAD 编程经验的工程师，都可以根据软件的提示迅速掌握诀窍，设计出自己想要的零件模型。

4.3.2 编程助手

新增的一个数控铣加工编程模块，其具有方便的代码编辑功能，简单易学，非常适合手工编程使用。同时，支持自动导入代码和手工编写的代码，其中包括宏程序代码的轨迹仿真，能够有效验证代码的正确性。支持多种系统代码的相互后置转换，实现加工程序在不同数控系统上的程序共享，还具有通信传输的功能，通过 RS 232 口可以实现数控系统与编程软件之间的代码互传。

多种粗、半精、精、补加工方式如下所示。

（1）提供 7 种粗加工方式：平面区域粗加工（2D）、区域粗加工、等高粗加工、扫描线粗加工、摆线粗加工、插铣粗加工、导动线（2.5 轴）粗加工。

（2）提供 14 种精加工方式：平面轮廓、轮廓导动、曲面轮廓、曲面区域、曲面参数线、轮廓线、投影线、等高线、导动、扫描线、限制线、浅平面、三维偏置、深腔侧壁多种精加工。

（3）提供 3 种补加工方式：等高线补加工、笔式清根补加工、区域补加工。

（4）提供 2 种槽加工方式：曲线式铣槽、扫描式铣槽。

（5）多轴加工主要包括两类：四轴加工包括四轴曲线、四轴平口面加工；五轴加工包括五轴等参数线、五轴侧铣、五轴曲线、五轴曲面区域、五轴 G01 钻孔、五轴定向、转四轴轨迹等加工功能。多轴加工针对叶轮、叶片类零件。除以上这些加工方法外，系统还提供专用的叶轮粗加工及叶轮精加工功能，可以实现对叶轮和叶片的整体加工。

（6）宏加工：提供倒圆角加工，根据给定的平面轮廓曲线，生成加工圆角的刀具轨迹和带有宏指令的加工代码。充分利用宏程序功能，使得倒圆角加工程序变得异常简单灵活。

（7）系统支持高速加工：可设定斜向切入和螺旋切入等接近和切入方式，拐角处可设定圆角过渡、轮廓与轮廓之间可通过圆弧或 S 字形方式来过渡形成光滑连接，生成光滑的刀具轨迹，有效地满足了高速加工对刀具轨迹形式的要求。

4.3.3 等高线加工

在零件特征栏中单击"加工管理"，在空白处右击后，在弹出的下拉菜单中选择"加工"→"粗加工"→"等高线粗工"命令，在弹出的"等高线粗加工"对话框中，单击"加工参数 1"标签，进入"加工参数 1"选项卡，"加工方向"选择"顺铣"，"Z 切入"选择"层高"，"层高"输入"0.3"。"XY 切入"选择"行距"，"行距"输入"7"，"切削模式"选择"环切"，"加工余量"输入"0.5"，如图 4-2 所示。

项目四 支撑座铣削编程加工训练

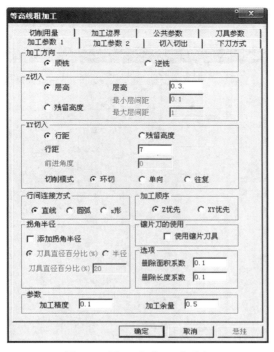

图 4-2 "加工参数 1"选项卡

单击"切入切出"标签，进入"切入切出"选项卡，在"方式"选项组中勾选"螺旋"，在"螺旋"选项组中输入"半径"为"5"，"螺距"为"0.3"，如图 4-3 所示。

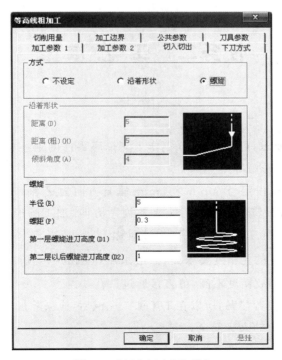

图 4-3 "切入切出"选项卡

单击"加工边界"标签,进入"加工边界"选项卡,在"相对于边界的刀具位置"选项组中选择"边界上",如图 4-4 所示。

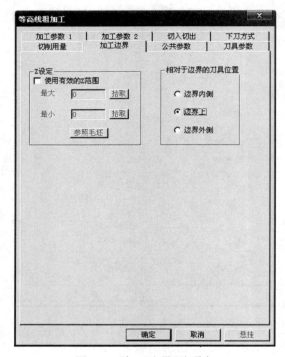

图 4-4 "加工边界"选项卡

最后单击"切削用量"标签,进入"切削用量"选项卡,合理设置加工参数。

任务二 工 艺 准 备

4.4 零件图分析

根据零件的使用要求,可以选择 45 钢作为支撑座零件的毛坯材料,毛坯下料尺寸定为 60mm×20mm×40mm 的方料。

如图 4-5 所示,可以设定与 60mm×40mm 垂直方向为 Z 轴,加工轮廓,再通过反向装夹加工厚度为 12mm 的精度尺寸,最后装夹左视图方向,加工圆形腔,保证精度与对称,反向装夹以右视图为基准加工沉头和通孔,保证对称。

在铣削加工时,$\phi15$ 的孔精度要求高,故应采用粗加工→半精加工→精加工顺序进行;$2\times\phi5.5$ 为螺栓安装孔,故精度不高,可直接通过钻孔获得。

注意,此零件与操作者接触,所以在工序加工完成后,应去除加工毛刺,保证锐角充分倒钝,以确保在使用过程中的人身安全。

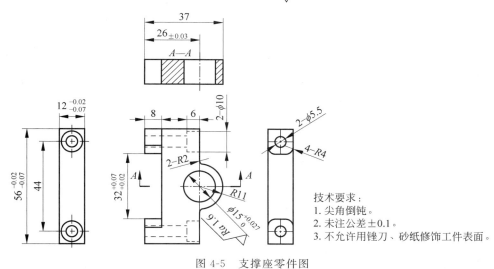

图 4-5 支撑座零件图

4.5 工艺设计

根据零件图分析,确定工艺过程,如表 4-1 所示。

表 4-1 工艺过程卡片

机械加工 工艺过程卡片		产品型号	XSB	零部件序号	XSB-04	第 1 页	
		产品名称	吸水泵	零部件名称	支撑座	共 1 页	
材料牌号	C45	毛坯规格	60mm×40mm ×20mm	毛坯质量	kg	数量	1
工序号	工序名	工序内容		工段	工艺装备	工时/min	
						准结	单件
5	备料	毛坯准备 60mm×40mm×20mm 方料		外购	锯床		
10	铣削	以毛坯面 60mm×40mm 为粗基准,精加工上表面		加工中心	平口钳	115	10
15	铣削	精加工侧壁,精加工圆形腔		加工中心	平口钳 游标卡尺		35
20	铣削	以 60mm×40mm 上边面为 Z 向基准定位,夹持 40mm×20mm 两个侧边,反向装夹,精加工平面		加工中心	平口钳 游标卡尺		15
25	铣削	以 60mm×20mm 面为 Z 向基准,夹持 60mm×40mm 平面,加工通孔和沉头		加工中心	平口钳 游标卡尺		25
30	铣削	以两个台阶做 Z 向定位,精加工 32mm×20mm 槽		加工中心	平口钳 游标卡尺		20
35	清理	清理工件,锐角倒钝		钳			5
40	检验	检验工件尺寸		检			5

本训练任务针对第 10～第 30 工序铣加工，进行工序设计，制订工序卡片，如表 4-2 所示。

表 4-2 铣加工工序卡片

机械加工工序卡片	产品型号	XSB	产品型号	XSB-04	第 1 页
	产品名称	吸水泵	产品名称	支撑座	共 1 页

工序号		10～30
工序名		铣制
材料		C45
设备		加工中心
设备型号		AVL650e
夹具		平口钳
量具		游标卡尺
准结工时		115min
单件工时		110min

技术要求：
1. 尖角倒钝。
2. 未注公差±0.1。
3. 不允许用锉刀、砂纸修饰工件表面。

工步	工步内容	刀具	S/(r/min)	F/(mm/r)	a_p/mm	a_e/mm	工步工时/min 机动	辅助
1	工件安装							5
2	60mm×40mm 平面加工	φ40 面铣刀	1500	800	1	20	5	
3	粗加工外轮廓，余量为 0.3mm	φ10 立铣刀	1500	400	3	5	10	
4	精加工外轮廓	φ4 立铣刀	1500	150	13	1	5	
5	粗加工圆形腔，余量为 0.1mm	φ10 立铣刀	1500	200	3	5	10	
6	精镗圆形腔	φ15 镗刀	400	80	0.5	15	10	
7	反向装夹，以 40mm×20mm 两面为精基准，以 60mm×40mm 加工平面为 Z 向基准							5
8	精加工 60mm×40mm 未加工平面	φ40 面铣刀	1500	400	3	20	10	
9	以 60mm×40mm 面为精基准，60mm×20mm 面为 Z 向基准定位							5
10	钻定心孔	φ8 定心钻	2500	400	2		5	
11	钻 φ5.5 通孔	φ5.5 麻花钻	600	150	10		10	
12	铣沉头	φ10 立铣刀	1500	300	6		5	

续表

工步	工步内容	刀具	S/ (r/min)	F/ (mm/r)	a_p/ mm	a_e/ mm	工步工时/min	
							机动	辅助
13	反向装夹,以60mm×40mm面为精基准,以两个台阶为Z轴定位基准							5
14	粗加工底部轮廓	φ10立铣刀	1500	600	1	5	10	
15	精加工底部轮廓	φ10立铣刀	1500	400			5	
16	拆卸、清理工件							5

任务三 软件编程训练

4.6 编程练习

4.6.1 创建加工坐标系及毛坯

首先创建加工坐标系,随后创建毛坯为60mm×40mm×20mm(见图4-6),将基准点移动,使毛坯包覆工件,并将工件填补,构造加工部分造型(见图4-7)。

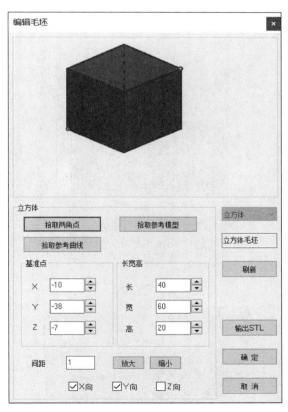

图4-6 创建毛坯

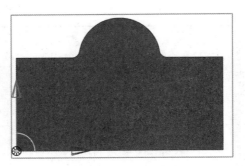

图 4-7　构造加工部分造型

4.6.2　加工刀具轨迹

加工刀具轨迹如图 4-8～图 4-11 所示。

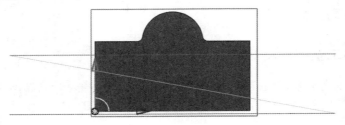

图 4-8　平面加工刀具轨迹

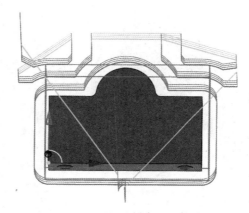

图 4-9　粗加工轮廓刀具轨迹

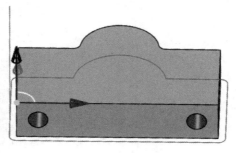

图 4-10　精加工轮廓刀具轨迹

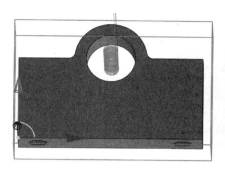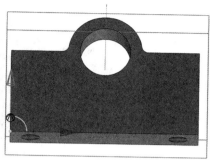

图 4-11　粗、精加工圆形腔刀具轨迹

注意：在圆形腔加工时，须选择加工边界；此圆形腔为与曲柄轴配合面，在加工时需要选择精镗加工。

4.6.3　反向装夹加工刀具轨迹

反向装夹加工刀具轨迹如图 4-12 所示。

图 4-12　反向装夹加工刀具轨迹

4.6.4　侧向装夹孔加工刀具轨迹

侧向装夹孔加工刀具轨迹如图 4-13 和图 4-14 所示。

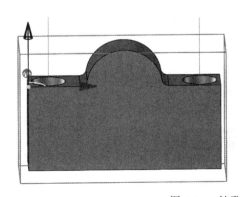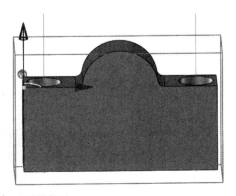

图 4-13　钻孔加工刀具轨迹

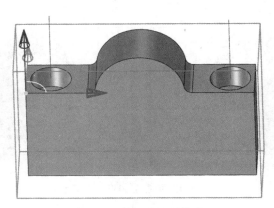

图 4-14　铣沉头刀具轨迹

4.6.5　底部不规则槽加工刀具轨迹

底部不规则槽加工包括粗加工和精加工，分别如图 4-15 和图 4-16 所示。

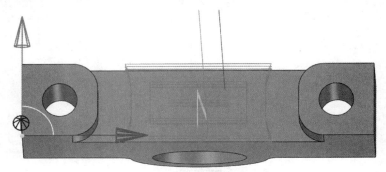

图 4-15　底部不规则槽轮廓粗加工刀具轨迹

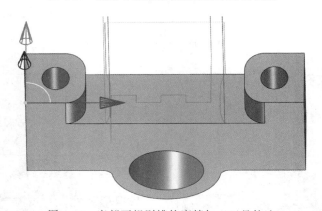

图 4-16　底部不规则槽轮廓精加工刀具轨迹

项 目 总 结

通过支撑座数控铣削加工，需要掌握数控平面加工、孔加工、等高线加工等主要加工方法。掌握机床配置加工参数、刀具轨迹检查和后置处理配置。

通过任务训练,养成良好的职业素养,培养正确的编程操作规范和基本安全素养,养成基本的机械加工质量和安全意识。

课 后 习 题

1. 填空题

(1) 制造工程师 2020 是基于_____平台全新开发的 CAD/CAM 一体化系统。

(2) 列举制造工程师软件实体造型中的造型方式:_____、_____、_____、_____、_____、_____等。

(3) 制造工程师 2020 提供曲面线裁剪和_____、_____、_____或选定曲面切矢的_____、多张曲面之间的拼接功能。

(4) 制造工程师 2020 新增的数控铣加工编程模块支持多种系统代码的_____,实现加工程序在不同数控系统上的程序共享。

(5) 制造工程师 2020 软件提供_____种精加工方式。

2. 判断题

(1) 制造工程师 2020 可以将二维的草图轮廓快速生成三维实体模型。　　(　　)

(2) 制造工程师 2020 相比于 CAXA 3D 实体设计 2020 新增了一个数控铣加工编程模块。　　(　　)

(3) 制造工程师 2020 提供 3 种槽加工方式。　　(　　)

(4) 在用轨迹法切削槽类零件时,槽两侧表面一面为顺铣、一面为逆铣,但两侧质量相同。　　(　　)

(5) 在工序加工完成后,应去除加工毛刺,保证锐角充分倒钝,以确保在使用过程中的人身安全。　　(　　)

3. 选择题

(1) 以下不是制造工程师 2020 的工作窗口中的是(　　)。
　　A. 管理树栏　　　B. 图素栏　　　C. 加工选项卡　　D. 绘图笔刷

(2) 以下不是制造工程师 2020 的 NURBS 曲面造型手段的是(　　)。
　　A. 扫描　　　　　B. 对称　　　　C. 旋转　　　　　D. 网格

(3) 制造工程师 2020 支持的模拟加工方式有(　　)。
　　A. 多轴加工　　　B. 宏加工　　　C. 高速加工　　　D. 以上均可

(4) 精加工常用指令有(　　)指令。
　　A. G70　　　　　B. G71　　　　　C. G73　　　　　D. 无

(5) 有关制造工程师 2020 错误的是(　　)。
　　A. 支持先进实用的轨迹参数化和批处理功能
　　B. 支持高速切削
　　C. 提供了知识加工功能、通用后置处理,还包含大量设计元素库
　　D. 操作复杂,涵盖范围小,加工方式有限

4. 简答题

(1) 简述平面铣和等高线铣削使用场景。

(2) 常见刀具有哪些?

(3) 如何通过工艺工序选定夹具和刀具?

自我学习检测评分表如表 4-3 所示。

表 4-3　自我学习检测评分表

项目	目标要求	分值	评分细则	得分	备注
学习关键知识点	(1) 了解 AVL650e 立式加工中心的结构及主要参数 (2) 理解铣削加工的特点 (3) 熟悉常用铣刀的分类,并能进行刀具的正确选择 (4) 了解数控编程基本思路 (5) 掌握数控铣削编程平面、圆形腔和孔的加工方法	25	理解与掌握		
工艺准备	(1) 能够正确识读基本平面类零件图 (2) 能够独立确定加工工艺路线,并正确填写工艺文件	25	理解与掌握		
编程训练	(1) 会正确选择相应的设备与用具 (2) 能够正确使用软件编程,并给出合理参数	50	(1) 理解与掌握 (2) 操作流程		

思政小课堂

项目五 底座铣削编程加工训练

➢ **思维导图**

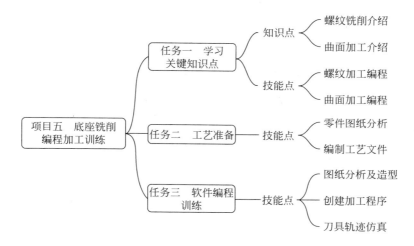

➢ **学习目标**

知识目标

(1) 了解三轴曲面加工。
(2) 了解球铣刀比立铣刀切削曲面质量更好的原因。

能力目标

(1) 掌握坐标系旋转的使用方法。
(2) 掌握螺纹加工指令。
(3) 掌握螺纹刀具的选择与使用方法。
(4) 能够独立确定加工工艺路线,并正确填写工艺文件。

素养目标

(1) 培养学生的科学探究精神和态度。
(2) 培养学生的工程意识。
(3) 培养学生的团队合作能力。

任务一 学习关键知识点

5.1 螺纹铣削加工

5.1.1 螺旋铣削内孔

1. 加工范围

孔径较大的盲孔或通孔,由于麻花钻加工太慢或不能加工,往往选择螺旋铣削的方式,而且由于该方式选择的刀具不带底刃,所以更适合小切深、高转速及大进给的加工情况。

2. 加工特点

螺旋铣削加工孔是建立在螺旋式下刀方法基础上的加工方法,螺旋铣孔时有一个特点:每螺旋铣削一周,刀具沿着 Z 轴方向移动一个下刀高度。

3. 说明

这种方法在螺旋铣削内孔时很有特色,其程序编写的实质就是将一个下刀高度作为螺旋线高度编写成一个子程序,通过循环调用该螺旋线子程序,完成整个孔的铣削加工。该方法加工孔不受铣刀规格等因素影响,所以在数控铣床和加工中心上应用广泛。

5.1.2 单刃螺纹铣刀加工螺纹

1. 加工范围

同一把螺纹铣刀既可以铣削左旋螺纹,又可以铣削右旋螺纹;既可以铣削内旋螺纹,又可以铣削外旋螺纹,同时,不受螺距和螺纹规格的影响。

2. 加工特点

单刃螺纹铣刀加工螺纹是建立在螺旋式下刀方法基础上的加工方式。铣螺纹的原理为:螺纹铣刀每铣一周,刀具在 Z 轴方向上运动一个导程(单线时为一个螺距)。

3. 说明

这种方法在螺纹铣削时很有特色,其程序编写的实质就是将一个导程的螺旋线编写成一个子程序,通过循环调用该螺旋线子程序,完成整个螺纹的铣削加工。该方法加工螺纹不受铣刀螺距和螺纹规格等参数的影响,所以在数控铣床和加工中心上应用广泛。

5.1.3 多刃螺纹铣刀加工螺纹

1. 加工范围

同一把螺纹铣刀既可以铣削左旋螺纹也可以铣削右旋螺纹,同时也可以铣削内外螺纹,主要用在生产效率高的场合。

2. 加工特点

每把铣刀对应一个值,该值为刀具圆角半径编程值,也就是在铣螺纹时的刀具半径补偿值。在铣削螺纹时,一般的加工深度可以一次加工完成,但是如果要求分多次铣削,则只需修改刀具半径补偿值就可以完成。

3. 说明

这种方法在螺纹铣削时效率非常高,程序编写也非常简单。其编程的实质是:螺纹铣刀在 XOY 平面导入(或导出)1/4 周时,正式加工螺纹 1 周;在 Z 轴方向导入(或导出)1/4 周时,刀具运动 1/4 个螺距;正式加工螺纹 1 周时,刀具在 Z 轴方向上运动 1 个螺距。通过保证多刃螺纹铣刀上的每个有效刀齿同时参与铣削,来完成整个螺纹的铣削加工。该方法加工螺纹的重点体现在选择铣刀的螺距上:当要求加工一定规格螺距的螺纹时,必须选择与其相对应的螺纹铣刀。同时,该方法受铣刀螺距和螺纹规格等因素的影响,但由于加工效率高,所以在数控铣床和加工中心上应用广泛。

5.2 曲面加工

三轴铣削加工可得到较好的弯曲近似曲面。在三轴铣削加工使用球头刀具时,以 X、Y、Z 方向的直线进给运动保证了刀具可切削到工件上任意位置的坐标点,但不能改变刀轴方向。刀轴上在这一点的实际切削速度为零,刀中心的排屑空间也很小。

曲面是 CAD 模型中的数学实体,可以准确表示标准几何对象(如平面、圆柱体、球体和圆环),以及雕刻自由形状的几何体。自由形式的几何体在设计领域有无数的应用。

曲面加工是一个多用途的功能,无论是以加工面积为导向,还是以操作为导向,其均可以完全适应需求和工作方法。可以在零件上定义加工区域,然后为每个区域分配一个操作;也可以将加工过程定义为一系列操作,每个操作都有一个加工区域。

5.2.1 三轴曲面加工的意义

三轴曲面加工是定义和管理数控程序的新一代产品。三轴曲面加工是一种基于三轴加工技术的三维几何加工技术。它特别适合模具和工具制造商,以及所有分支机构和工业各级原型制造商的需要。

三轴曲面加工提供易于学习和使用,且面向车间的三轴制造的刀具轨迹定义。三轴曲面加工是基于行业公认的、领先的技术。

薄壁曲面零件广泛应用于工业生产中,其形状与精度是保证其可用性的基本要求。由于薄壁曲面零件的刚性低,因此,在加工过程中切削力成为加工形变的敏感因素。而三轴高速铣削与传统铣削相比,具有明显的切削力小的特性,为钛合金等难加工材料制成的薄壁曲面零件的加工,提供了一种有效的方法。

三轴数控加工中心因其操作简单而应用广泛。在加工曲面时,三轴数控加工中心使用插补线段来拟合曲面。加工表面的质量受插补线段长度的影响:在直线段的连接处形成锐角,锐角的出现会导致应力集中增加。

5.2.2 三轴曲面加工方法

3D 零件的几何形状和所需的表面粗糙度,在确定用于任何给定零件的刀具轨迹方面都起着关键作用。三轴曲面加工的目标是计算零件表面上和沿着零件表面的路径,以便刀具在切削时可以遵循。通常,三轴刀具轨迹被投影到下面的表面上。

三轴数控车削主要是把工件旋转后,用刀具将工件切割成所需的形状。当刀具在平行旋转轴线上运动时,可获得内外圆柱面;而圆锥表面的形成,则是刀具与轴心相交的斜线运动;旋转曲面的形成,是仿形数控车床控制工具沿曲线进给;另外一种旋转曲面的形成方式,则是采用成型车刀,横向进给。此外,螺纹表面、端平面和偏心轴的加工也可以通过三轴数控车削来实现。

三轴曲面切削主要包括仿形铣削、数控铣削及特殊加工方法。仿形铣削利用零件原型,对球头的仿形头进行一定压力的加工,使其与零件原型曲面接触。加工过程中,将仿形头的运动动能转化为电感,通过控制铣床三轴的运动,形成球头与仿形半径相结合的铣刀。CNC 技术的出现,为曲面加工提供了更为有效的手段。

任务二　工　艺　准　备

5.3　零件图分析

如图 5-1 所示,根据零件的使用要求,选择 45 钢为底座零件的毛坯材料,毛坯下料尺寸定为 95mm×95mm×30mm 的方料。

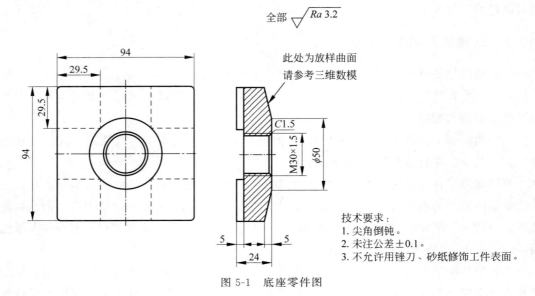

图 5-1　底座零件图

零件 M30 为关键尺寸,需要与其他零件配合;零件外形无配合要求,有形状即可,所以粗加工即可;曲面处需要精加工。

注意:此零件为操作者接触件,所以在工序加工完成后,应去除加工毛刺,保证锐角充分倒钝以确保在使用过程中的人身安全。

5.4 工艺设计

根据零件图分析,确定工艺过程,如表 5-1 所示。

表 5-1 工艺过程卡片

机械加工工艺过程卡片		产品型号	XSB	零部件序号	XSB-05	第 1 页	
^^		产品名称	吸水泵	零部件名称	底座	共 1 页	
材料牌号	C45	毛坯规格	95mm×95mm×30mm	毛坯质量	kg	数量	1
工序号	工序名	工序内容		工段	工艺装备	工时/min	
^^	^^	^^		^^	^^	准结	单件
5	备料	毛坯准备 95mm×95mm×30mm 的方料		外购	锯床		
10	铣削	以毛坯面 95mm×95mm 为粗基准,精加工上表面		加工中心	平口钳	160	5
15	铣削	精加工曲面		加工中心	平口钳		35
20	铣削	精加工 95mm×95mm 侧边		加工中心	平口钳 游标卡尺		25
25	铣削	钻螺纹底孔,铣削螺纹		加工中心	平口钳 螺纹通止规		30
30	铣削	反向装夹,铣平面		加工中心	平口钳 游标卡尺		20
35	铣削	精加工十字槽		加工中心	平口钳 游标卡尺		30
40	清理	清理工件,锐角倒钝		钳			5
45	检验	检验工件尺寸		检			5

本训练任务针对第 10～第 35 工序铣加工,进行工序设计,制作工序卡片,详细编制工艺卡,如表 5-2 所示。

表 5-2 铣加工工序卡片

机械加工工序卡片	产品型号	XSB	零部件序号	XSB-05	第 1 页
	产品名称	吸水泵	零部件名称	底座	共 1 页

全部 $\sqrt{Ra\ 3.2}$

工序号	10～35
工序名	铣削
材料	C45
设备	加工中心
设备型号	AVL650e
夹具	平口钳
量具	游标卡尺
准结工时	160mim
单件工时	155min

技术要求：
1. 尖角倒钝。
2. 未注公差±0.1。
3. 不允许用锉刀、砂纸修饰工件表面。

工步	工步内容	刀具	$S/$ (r/min)	$F/$ (mm/r)	$a_p/$ mm	$a_e/$ mm	工步工时/min 机动	工步工时/min 辅助
1	工件安装							5
2	铣平面	ϕ40 面铣刀	1500	400	0.5	20	5	
3	粗加工曲面	ϕ40 面铣刀	1500	800	0.3	20	10	
4	精加工曲面	R5 球头铣刀	3000	800	0.1	5	25	
5	粗加工外轮廓	ϕ10 立铣刀	1500	600	1.5	5	20	
6	精加工外轮廓	ϕ10 立铣刀	1500	400	25	5	5	
7	钻孔定心	ϕ8 定心钻	2500	200	2		5	
8	钻孔	ϕ28.5 麻花钻	800	100	30		10	
9	铣螺纹	ϕ16 螺纹铣刀	800	60			10	
10	倒斜角	ϕ8 定心钻	2500	400	1.5	1.5	5	
11	反向装夹，以 94mm×94mm 面为精基准，以第 2 步加工平面为 Z 向基准							5
12	粗加工平面	ϕ40 面铣刀	1500	800	2	20	10	
13	精加工平面	ϕ40 面铣刀	1500	400	0.3	20	5	
14	粗加工底部十字槽	ϕ10 立铣刀	1500	600	1	5	20	
15	精加工底部轮廓	ϕ10 立铣刀	1500	400	5	5	10	
16	拆卸、清理工件							5

任务三　软件编程训练

5.5　编程练习

5.5.1　创建毛坯及加工坐标系

如图 5-2 所示，创建毛坯及加工坐标系。

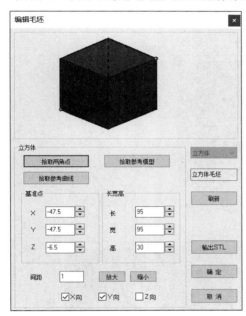
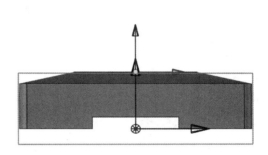

图 5-2　创建毛坯及加工坐标系

5.5.2　加工刀具轨迹

加工刀具轨迹如图 5-3～图 5-7 所示，包括平面精加工、曲面粗加工和曲面精加工、外轮廓粗加工和精加工、螺纹铣削及倒角。此处使用的平面轮廓加工，需要构造草图轮廓，选取模型边界，另外为使刀具轨迹优化，需要自定义进退刀点。在加工螺纹孔倒角时，需要注意此处构造为螺纹底孔圆。

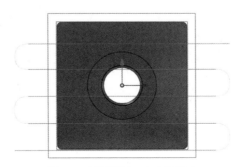

图 5-3　平面精加工刀具轨迹

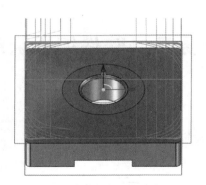

图 5-4 曲面粗加工刀具轨迹

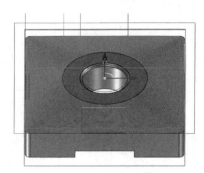

图 5-5 曲面精加工刀具轨迹

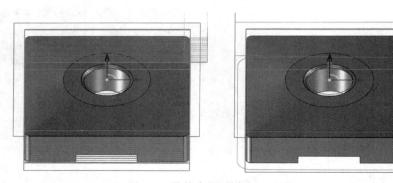

图 5-6 外轮廓粗、精加工刀具轨迹

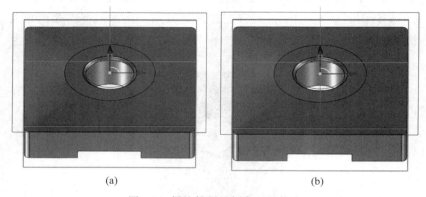

图 5-7 螺纹铣削及倒角刀具轨迹

(a) 定心钻定心刀具轨迹；(b) 麻花钻螺纹底孔刀具轨迹；(c) 螺纹铣削加工刀具轨迹；(d) 螺纹孔倒角刀具轨迹

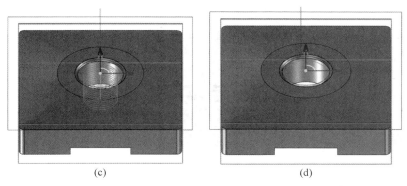

(c)   (d)

图 5-7 （续）

5.5.3 反向装夹加工刀具轨迹

反向装夹加工刀具轨迹如图 5-8 和图 5-9 所示，包括反向装夹平面加工、底部十字槽的粗加工和精加工。

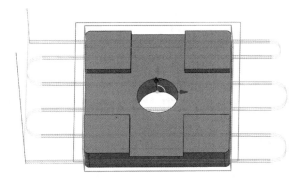

图 5-8 反向装夹平面加工刀具轨迹

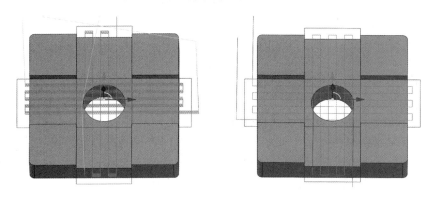

图 5-9 底部十字槽粗、精加工刀具轨迹

项 目 总 结

通过底座数控铣削加工，需要掌握数控铣削刀具的选用和设置，以及平面加工、圆形腔加工、螺纹加工及孔加工的程序编写，并合理配置加工参数，对刀具轨迹进行检查与优化。

最后利用后置处理生成代码,并且学会检查代码命令。

通过任务训练,养成良好的职业素养,培养正确的加工中心安全操作规范,养成基本的机械加工质量意识。

课后习题

1. 填空题

(1) 螺旋铣削内孔的两种常用刀具是_____、_____。

(2) 为得到较好的弯曲度,曲面加工可采用_____。

(3) 三轴曲面加工是一种基于_____的三维几何加工技术。

(4) 螺纹孔加工的常用指令有_____。

(5) 螺纹孔加工的常用指令中 F 代码代表的是_____。

2. 判断题

(1) 单刃螺纹铣刀加工螺纹原理:螺纹铣刀每铣一周,刀具在 Z 轴方向上运动一个导程。( )

(2) 多刃螺纹铣刀在加工螺纹时效率非常低,程序编写也非常复杂。( )

(3) 在三轴铣削加工曲面时,刀轴上在这一点的实际切削速度为零,刀中心的排屑空间也很小。( )

(4) 三轴曲面加工特别适合模具和工具制造商,以及所有分支机构和工业各级原型制造商的需要。( )

(5) 三轴高速铣削与传统铣削相比,具有明显的切削力大的特性。( )

3. 选择题

(1) 用 FANUC 系统指令"G92 X(U) Z(W) F;"加工双线螺纹,则该指令中的"F"是指( )。

  A. 螺纹螺距   B. 螺纹导程

  C. 每分钟进给量   D. 每转进给量

(2) 在下列选项中( )指令不被用于螺纹加工。

  A. G32   B. G92   C. G76   D. G81

(3) 多刃螺纹铣刀加工螺纹的刀具选择要求( )。

  A. 匹配螺纹规格   B. 匹配螺纹螺距

  C. 每个有效刀齿同时参与铣削 D. 以上均可

(4) 三轴曲面切削主要包括( )。

  A. 仿形铣削 B. 数控铣削 C. 特殊加工方法 D. 以上均可

(5) 有关三轴曲面加工错误的是( )。

  A. 定义和管理数控程序的新一代产品

  B. 形状精度可靠

  C. 即时周期更新

  D. 操作复杂、涵盖范围小、加工方式有限

4. 简答题

(1) 简述底座曲面加工需要配置的参数。

(2) 证明在使用球头铣刀切削斜面时,加工工件表面质量比使用立铣刀效果好。

自我学习检测评分表如表 5-3 所示。

表 5-3 自我学习检测评分表

项目	目标要求	分值	评分细则	得分	备注
学习关键知识点	(1) 掌握坐标系旋转指令的使用方法 (2) 掌握螺纹和曲面的编程方法	20	理解与掌握		
工艺准备	(1) 能够正确识读零件图 (2) 能够独立确定加工工艺路线,并正确填写工艺文件 (3) 能够根据工序加工工艺,编写正确的加工程序	30	理解与掌握		
编程训练	(1) 能够根据零件结构特点和精度,合理选用加工方案 (2) 掌握底座铣削的工艺流程 (3) 能够正确根据零件情况调整加工参数	50	(1) 理解与掌握 (2) 操作流程		

思政小课堂

项目六　曲柄连杆铣削编程加工训练

➢ 思维导图

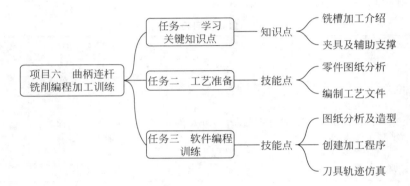

➢ 学习目标

知识目标

(1) 了解铣槽加工方法。
(2) 了解异形零件工装设计及加工。

能力目标

(1) 掌握铣槽指令的使用方法。
(2) 掌握运用工装夹具加工不规则零件。

素养目标

(1) 培养学生的科学探究精神和态度。
(2) 培养学生的工程意识。
(3) 培养学生的团队合作能力。

任务一　学习关键知识点

6.1　切槽铣刀

1. 立铣刀

立铣刀(见图 6-1)用于加工沟槽和台阶面等,刀齿在圆周和端面上;一般在工作时不能沿轴向进给,但当立铣刀上有通过中心的端齿时,可轴向进给。切削刃有双刃、三刃、四刃,直径 $\phi 2 \sim \phi 15 \mathrm{mm}$,大量应用于切入式铣削、高精度沟槽加工等。

图 6-1　立铣刀

2. 三面刃铣刀

三面刃铣刀(见图 6-2)用于加工各种沟槽和台阶面,其两侧面和圆周上均有刀齿。在切削直角形的角落或沟槽时所使用的三面刃铣刀,在构造上可分为切削刃相互交错的错齿形三面刃铣刀,以及切削刃平行排列的并齿形三面刃铣刀。并齿形三面刃铣刀是最常用的,而错齿形三面刃铣刀则用于钢材的沟槽加工。在批量生产中,经常使用大量的可转位刀具,其中包括半三面刃铣刀和全三面刃铣刀。全三面刃铣刀多用于沟槽的加工。

图 6-2　三面刃铣刀

3. 锯片铣刀

锯片铣刀(见图 6-3)用于加工深槽和切断工件,其圆周上有较多的刀齿。为了减少在铣切时产生的摩擦,刀齿两侧有 $15'\sim 1°$ 的副偏角。锯片铣刀的特点:可使用磨齿机重复多次重磨刃齿,研磨后的锯片铣刀与新锯片铣刀寿命相同。

4. T 形槽铣刀

T 形槽铣刀(见图 6-4)用来铣 T 形槽。T 形槽铣刀可以分为锥柄 T 形槽铣刀和直柄 T 形槽铣刀,可用于加工各种机械台面或其他结构体上的 T 形槽,是加工 T 形槽的专用工具,可一次铣出精度达到要求的 T 形槽。铣刀端刃有合适的切削角度,刀齿按斜齿、错齿设计,切削平稳、切削力小。

图 6-3　锯片铣刀　　　　　图 6-4　T形槽铣刀

6.2　夹具及辅助支撑

使工件在机床上相对刀具占有正确的位置的过程称为定位；克服在切削过程中工件所受到的外力作用，保持工件的准确位置的过程称为夹紧。以上两者综合称为装夹，装夹时使用的工艺装备称为机床夹具。

机床夹具按通用化程度可分为两大类。

1. 通用夹具

只需调整或更换少量零件，就可用于装夹不同的工件，如三爪卡盘、四爪卡盘、顶尖、平口钳等。

通用夹具的结构复杂，可适用于大批量生产，也可适用于单件小批量生产。

2. 专用夹具

专用夹具是专门为某工件工序设计和制造的专用夹具，其结构简单、紧凑、操作迅速、方便，专用夹具适用于产品固定的成批或大量生产。

任务二　工　艺　准　备

6.3　零件图分析

如图 6-5 所示，根据零件的使用要求，选择 45 钢为曲柄连杆零件的毛坯材料，毛坯下料尺寸定为 65mm×25mm×25mm 的方料。其中，一种加工方案为铣床加工 $\phi 8$ 精度的轴和端面，精加工 $\phi 8$ 的孔，铣出轮廓。反向装夹，三爪卡盘一夹一撑，加工 $\phi 16$ 圆和平面。

注意：此零件为手动操作件，所以在工序加工完成后，应去除加工毛刺，保证锐角充分倒钝，以确保在使用过程中的人身安全。

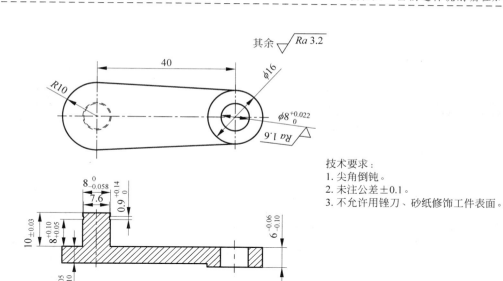

图 6-5 曲柄连杆零件图

6.4 工艺设计

根据零件图分析,确定工艺过程,如表 6-1 所示。

表 6-1 工艺过程卡片

机械加工工艺过程卡片		产品型号	XSB	零部件序号	XSB-06	第 1 页	
		产品名称	吸水泵	零部件名称	曲柄连杆	共 1 页	
材料牌号	C45	毛坯规格	65mm×25mm×25mm	毛坯质量	kg	数量	1
工序号	工序名	工序内容		工段	工艺装备	工时/min	
						准结	单件
5	备料	毛坯准备 65mm×25mm×25mm 的方料		外购	锯床		
10	铣削	以毛坯面 65mm×25mm 为粗基准,精加工上表面		加工中心	平口钳	130	5
15	铣削	精加工圆形凸台		加工中心	平口钳		30
20	铣削	精加工外轮廓		加工中心	平口钳 游标卡尺		20
25	铣削	铣槽		加工中心	平口钳		5
30	铣削	精加工孔		加工中心	平口钳 游标卡尺		15
35	铣削	反向装夹,精加工平面		加工中心	平口钳 三爪卡盘		15
40	铣削	精加工圆形凸台		加工中心	平口钳 游标卡尺		20
45	清理	清理工件,锐角倒钝		钳			5
50	检验	检验工件尺寸		检			5

本训练任务针对第10～第40工序铣加工,进行工序设计,制订工序卡片,详细编制工艺卡,如表6-2所示。

表6-2 铣加工工序卡片

机械加工工序卡片	产品型号	XSB	零部件序号	XSB-06	第1页
	产品名称	吸水泵	零部件名称	曲柄连杆	共1页

工序号	10～40
工序名	铣削
材料	C45
设备	加工中心
设备型号	AVL650e
夹具	平口钳
	三爪卡盘
量具	游标卡尺
准结工时	130min
单件工时	125min

其余 $\sqrt{Ra\ 3.2}$

技术要求:
1. 尖角倒钝。
2. 未注公差±0.1。
3. 不允许用锉刀、砂纸修饰工件表面。

工步	工步内容	刀具	S/(r/min)	F/(mm/r)	a_p/mm	a_e/mm	工步工时/min 机动	辅助
1	工件安装							5
2	铣平面	φ40面铣刀	1500	400	0.5	20	5	
3	粗加工圆形台和平面	φ40面铣刀	1500	800	0.3	20	20	
4	精加工圆形台和平面	φ10立铣刀	1500	600	0.1	5	10	
5	粗加工外轮廓	φ40面铣刀	1500	600	2	20	15	
6	精加工外轮廓	φ10立铣刀	1500	400	25	5	5	
7	切卡簧槽	φ20×0.9槽铣刀	1000	100			5	
8	钻孔定心	φ8定心钻	2500	200	2		5	
9	钻孔	φ7.8麻花钻	800	80	8		5	
10	铰孔	φ8铰刀	600	40			5	
11	反向装夹,三爪卡盘定位,平口钳支撑悬臂							5
12	粗加工平面	φ40面铣刀	1500	800	2	20	10	
13	精加工平面	φ40面铣刀	1500	400	0.3	20	5	
14	粗加工圆形台及平面	φ40立铣刀	1500	600	1	20	10	
15	精加工圆形台及平面	φ10立铣刀	1500	400	1	5	10	
16	拆卸、清理工件							5

任务三 软件编程训练

6.5 编程练习

6.5.1 创建毛坯及加工坐标系

首先创建毛坯,随后创建加工坐标系(见图 6-6)。

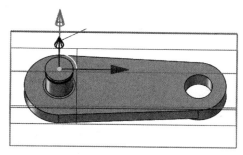

图 6-6 创建加工坐标系

6.5.2 加工刀具轨迹

加工刀具轨迹如图 6-7～图 6-11 所示,包括平面加工、圆形台及平面的粗加工和精加工、铣卡簧槽加工、外轮廓粗加工和精加工及孔加工。

注意:在铣卡簧槽加工时,此处槽轮廓圆小于圆形台的直径,需进行避让。

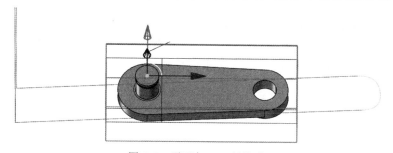

图 6-7 平面加工刀具轨迹

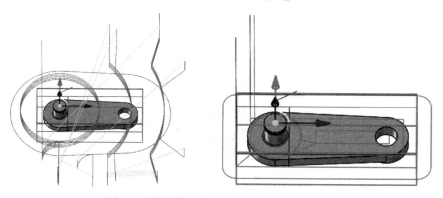

图 6-8 圆形台及平面的粗、精加工刀具轨迹

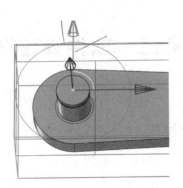

图 6-9　铣卡簧槽加工刀具轨迹

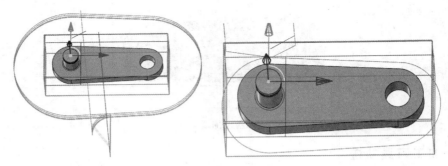

图 6-10　外轮廓粗、精加工刀具轨迹

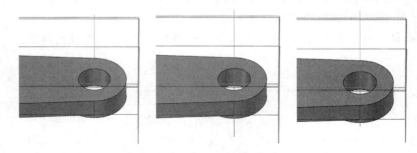

图 6-11　孔加工刀具轨迹

6.5.3　反向装夹加工刀具轨迹

反向装夹加工刀具轨迹如图 6-12 和图 6-13 所示，包括平面和轮廓的粗、精加工。

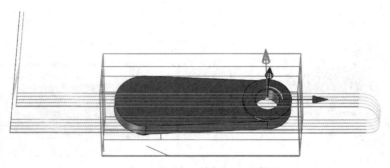

图 6-12　平面粗、精加工刀具轨迹

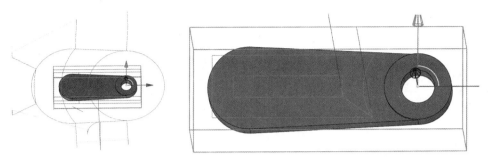

图 6-13 轮廓粗、精加工刀具轨迹

项 目 总 结

通过曲柄连杆加工编程，掌握切槽铣刀和夹具的灵活运用。

加工方法并不唯一，需根据现有条件和图纸精度合理选用机床，尽可能减少装夹次数，保证工时尽可能缩短。

通过任务训练，养成良好的职业素养，培养正确的加工中心安全操作规范，养成基本的机械加工质量意识。

课 后 习 题

1. 填空题

(1) 切槽铣刀的种类有_____、_____、_____、_____。

(2) T形槽铣刀可以分为_____T形槽铣刀和_____T形槽铣刀。

(3) 机床夹具可分为两大类，分别是_____、_____。

(4) 专用夹具具有_____、_____、_____等优点。

2. 判断题

(1) 一般来说立铣刀能沿轴向给进。　　　　　　　　　　　　　　　　　(　　)

(2) 三面刃铣刀中并齿形三面刃铣刀是最常用的，而错齿形三面刃铣刀则用于钢材的沟槽加工。　　　　　　　　　　　　　　　　　　　　　　　　　　　　(　　)

(3) 锯片铣刀用于加工深槽和切断工件，其圆周上有较多的刀齿，且可重复磨刃、延长寿命。　　　　　　　　　　　　　　　　　　　　　　　　　　　　　(　　)

(4) T形槽铣刀是加工T形槽的专用工具，但其切削力过大，经常导致震刀。(　　)

(5) 相比于通用夹具，专用夹具更适合产品固定的成批或大量生产。　　(　　)

3. 简答题

通过所学知识指定曲柄连杆零件加工的另一种方案。

自我学习检测评分表如表6-3所示。

表6-3 自我学习检测评分表

项目	目标要求	分值	评分细则	得分	备注
学习关键知识点	(1) 了解零件基本构成 (2) 掌握多种加工方法 (3) 设计新的加工方案	25	理解与掌握		
工艺准备	(1) 能够正确识读零件图 (2) 能够独立确定加工工艺路线 (3) 能够根据工序加工工艺,编写正确的加工程序	25	理解与掌握		
编程训练	能够根据工艺方案,编写程序并执行加工	50	(1) 理解与掌握 (2) 操作流程		

思政小课堂

Projects Guidance

CNC programming is an important work in CNC machining. The process from a part drawing to obtaining qualified a CNC machining program is CNC programming. Taking the suction pump model (Figure 0-1) as an example, this book uses the CAXA CAM CNC lathe (hereinafter referred to as the CNC lathe) and the CAXA CAM manufacturing engineer (hereinafter referred to as the manufacturing engineer) programming software to explain the process preparation, software operation and post-processing code generation from easy to difficult.

Figure 0-1 The suction pump model

Note: The software version is 2020, which will not be explained later. In addition, it assumed that the drawing ability has already been mastered before learning this book.

As shown in Figure 0-2, the suction pump mainly includes nine typical parts, which are involved in CNC turning, CNC milling and combined machining schemes. Taking the first six parts as the main line, the software programming method of the part characteristics of shaft, hole, thread, groove and die cavity is developed one by one. Each

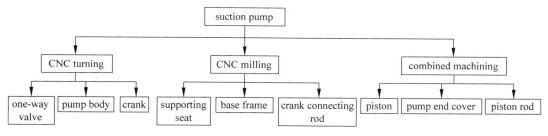

Figure 0-2 Typical project

task involves the necessary learning of the specialized knowledge and skills to complete the machining of parts, such as equipment, fixtures, cutters, basic instructions and machine operations. According to the technological design of CNC machining, targeted learning and training are carried out to master the technological design, CNC programming and machine operation of CNC turning and milling (The combined machining is arranged for the final practice and will not be repeated in the book).

Project 1 Programming and Machining Training for One-way Valve Turning

➢ Mind map

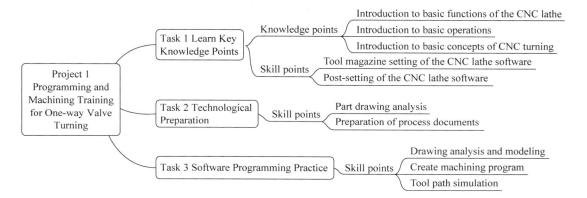

➢ Learning objectives

Knowledge objectives

(1) Understand the software window of CAXA CAM CNC lathe.

(2) Understand the pixel processing of drawings in the CNC lathe software.

(3) Understand the selection of machining parameters and the simulation test of tool path.

Ability objectives

(1) Analyze the drawings and determine the parts to be processed.

(2) Use graphic software to model the processing part.

(3) According to the processing conditions, select the appropriate processing parameters to generate the processing path.

(4) Track simulation test.

(5) Configure the machine tool, generate G codes, and send them to the machine tool for processing.

Literacy goals

(1) Cultivate students' enthusiasm for learning.

(2) Cultivate students' hands-on ability.

(3) Establish students' independent thinking ability.

➤ Task introduction

The machining process shall be formulated according to the requirements of the part drawing shown in Figure 1-1. Use the CNC lathe software to write the machining program and complete the machining of the one-way valve. As a typical shaft part, this part is made of 45 steel, and its surface is required to be smooth without scratches. It is not allowed to trim the surface of the part with abrasive paper or file.

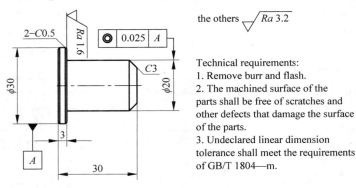

Figure 1-1　The part drawing of the one-way valve

Task 1　Learn Key Knowledge Points

1.1　Introduction to basic functions of the CNC lathe

CNC lathe software is a CNC lathe machining programming and two-dimensional graphic design software developed on a new CNC machining platform. CNC lathe software has powerful drawing function of CAD software and perfect external data interface. It can draw any complex graphics and exchange data with other systems through DXF, IGES and other data interfaces. The CNC lathe software provides powerful and concise path generation means, which can generate tool paths of various complex graphics according to machining requirements. The general post-processing module enables CNC lathe software to meet the code format of various machine tools, output G code, and verify and process the generated NC code. At the same time, it also provides anti-reading of NC code files and verification and simulation of tool paths.

The 2020 version of CNC lathe software is built based on the two-dimensional platform CAXA electronic drawing board, with changes in operation interface and operation style. Figure 1-2 shows the menu bar of the 2020 CNC lathe.

According to the processes commonly used in turning, such as roughing, finishing, grooving, and threading, professional functions of rough turning, finish turning, groove turning and three different thread machining have been developed in the 2020 CNC lathe

Figure 1-2　The menu bar of the 2020 CNC lathe software

software, as shown in Figure 1-3.

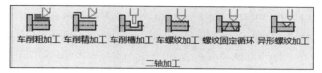

Figure 1-3　Two-axis machining option group

In addition to the standard two-axis machining function, the functions of equal section machining, radial direction and end face drilling, embedded and open key way machining are also developed for the turning center with C-axis, as shown in Figure 1-4.

Figure 1-4　C-axis machining option group

These functions can be used to easily generate the tool path, and the correctness of the tool path can be verified through the simulation function. After the verification is completed, the NC code file required by the machine tool can be converted through post-processing (Figure 1-5). The code file can be transferred to the machine tool for processing through the communication port in the CNC lathe software. The system has built-in some general post-processing files, such as FANUC, SIEMENS.

Figure 1-5　Post-processing option group

(1) The open post setting function allows users to customize the post setting according to the machine tool of the enterprise. It allows users to automatically generate code files that conform to the special machine tool according to the custom code of the special machine tool for processing.

(2) It supports the processing of small memory machine tool system, and supports to automatically segment and output the large programs.

(3) According to the requirements of the CNC system, if output the line number and if the line number is automatically filled in, and the programming method can be incremental or absolute.

(4) The coordinate output format can be defined to the decimal and integer digits.

(5) The arc output mode is set with the meaning of I, G, K and R modes.

The management tree bar (Figure 1-6) intuitively displays the tool, path, code, and other information of the current document in the form of a tree diagram, and provides many operation functions to facilitate the user to execute various commands related to the CNC lathe. The management tree box is located on the left side of the drawing area by default. Users can freely drag it to the desired location or hide it. The management tree bar has a machining master node, under which there are three sub-nodes: tool magazine, path and code, which are used to display and manage tool information, path information and G code information respectively.

Figure 1-6　The management tree

Integrated tool magazine management function, including the management of contour turning tools, grooving tools, threading tools, and drilling tools. It is convenient for users to obtain tool information from the tool magazine and maintain the tool magazine.

In the 2020 CNC lathe software, it is very convenient to create a tool by directly using the creating tool dialog box or using the "creating tools" in the right-click menu key and then left click "creat tools" command in the dropdown list. Moreover, the cutting parameters used by this tool are related to geometric parameters. The cutting elements of the tool are called at the same time when the tool is called during machining, saving the calling time for re-setting parameters (Figure 1-7).

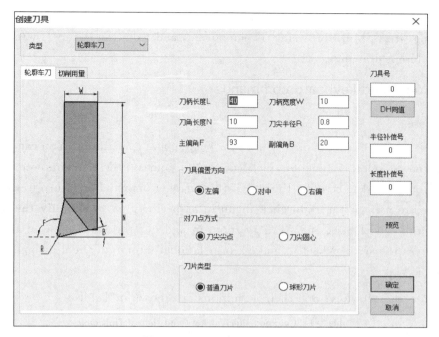

Figure 1-7 Tool management

1.2 Introduction to basic operations

1.2.1 Window layout

The window layout of CNC lathe software is shown in Figure 1-8. Tabs. All function commands can be found in the tab area.

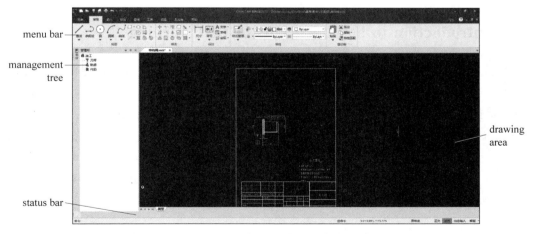

Figure 1-8 Window layout of CNC lathe software

Management tree. All tools, CNC lathe machining path and G code information will be recorded and displayed on the management tree.

Run menu. It refers to tab function run options and operation command prompt.

Drawing area. Multi-browsing is supported, and you can freely switch and edit between different drawings.

1.2.2 Mouse and keyboard commands

1. Mouse

The left mouse button is mainly used for picking and confirming. You can use the left mouse button to select a function, pick coordinate points, pick elements, etc. The right mouse button is used to confirm the end of the current command or return to the previous command, or to call up the right-click menu in a specific area to simplify the operation. Drag the scroll wheel back and forth to zoom in and out the view angle of the drawing area, and press the scroll wheel to make a translational motion of the drawing area.

2. Keyboard

Common shortcut keys and their definitions are shown in Table 1-1.

Table 1-1 Common shortcut keys and their definitions

shortcut key	definition	shortcut key	definition	shortcut key	definition
F1	Help files	F3	Show all	F5	Coordinate system switching
F6	Capture mode toggles	F7	3D view navigation switch	F8	Quadrature mode switch
F9	Window switching	Delete	Delete	Ctrl+P	Print

The remaining shortcut keys are consistent with common software, such as shortcut key for copying Ctrl+C.

1.3 Introduction to basic concepts of CNC turning

The process of machining with CNC lathe software is as follows. Firstly, analyze the drawings, determine the parts that need to be processed by CNC, and use graphic software to shape the parts that need to be processed. Then, according to the machining conditions, select the appropriate parameters to generate the tool path (including rough machining, semi-finish machining, finish machining path), and carry out the tool path simulation test. Finally, configure the machine tool, generate G code, and send it to the machine tool for processing.

1.3.1 Two-axis machining

In CNC lathe software, the Z axis of the machine coordinate system is the X axis of the absolute coordinate system, and the plane graphics are uniformly projected onto the XOY plane of the absolute coordinate system, as shown in Figure 1-9.

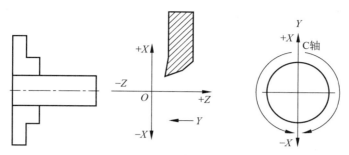

Figure 1-9 Two-axis machining

In general, it is necessary to establish a machining coordinate system, which forms a certain relationship with the coordinate system of the machine tool.

1.3.2 Contour

For the CNC lathe software, the contour is a collection of a series of end-to-end curves, as shown in Figure 1-10.

1.3.3 Blank contour

For rough machining, the machining range needs to be defined, namely, the blank contour as shown in Figure 1-11.

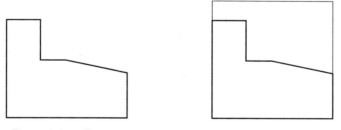

Figure 1-10 Contour   Figure 1-11 Blank contour

When interactively specifying the graphics to be processed in CNC programming, users are often required to specify the contour of the blank to define the surface to be processed or the blank itself to be processed.

Note: If the blank contour is used to define the machined surface, the specified contour is required to be closed. If the blank contour itself is processed, the blank contour can also not be closed.

1.4 Tool magazine settings of the CNC lathe software

Before processing with CNC lathe software, the tool, CNC system and machine tool need to be set up, which will directly affect the machining trajectory and the generated G code. This chapter will introduce these settings in detail.

This function of tool magazine management defines and determines the relevant data of the tool, so that the user can obtain the tool information from the tool magazine and maintain the tool magazine. The tool magazine management function includes the management of four tool types: contour turning tools, grooving tools, thread turning tools and drilling tools.

Operation method:

(1) Select menu bar→Create Tool→CNC lathe, and pop up the "Create Tool" dialog box, where users can add new tools according to their needs. The newly created tool list will be displayed under the tool magazine node of the management tree on the left side of the drawing area.

(2) Double-click the tool node under the tool magazine node to open the tool editing dialog box to change the tool parameters.

(3) Select the Export Tools command from the drop-down list that pops up after right-clicking the tool magazine node to save the information of all tools to a file.

(4) Select the Import Tools command from the dropdown list that pops up after right-clicking on the tool magazine node to read all the tool information saved in the file into the document and add it to the tool magazine node.

(5) It should be pointed out that all kinds of tools in the tool magazine are only abstract descriptions of the same type of tools, but not detailed tool magazine conforming to national standards or other standards. Therefore, only some parameters that have an impact on path generation are listed, while other tool parameters related to specific machining processes are not listed. For example, all kinds of external contour, internal contour and end face rough and fine turning tools are classified as contour turning tools, which has no effect on track generation.

1.4.1　Contour turning tools

The functional interface of contour turning tool tab is shown in Figure 1-12, and the specific parameters that need to be configured are as follows.

(1) Tool No.: the serial number of the tool, which is used for the automatic tool change command of post-processing. The tool number is unique and corresponds to the tool magazine of the machine tool.

(2) Radius compensation No.: the serial number of the tool radius compensation value, whose value corresponds to the database of the machine tool.

(3) Tool shank length: the length of the clamping section of the tool.

(4) Tool shank width: the width of the clamping section of the tool.

(5) Tool angle length: the length of the cutting section of the tool.

(6) Nose radius: the radius of the arc used for cutting in the tool tip part.

(7) Main deflection angle: the included angle between the front edge of the tool and the rotation axis of the workpiece.

(8) Secondary deflection angle: the included angle between the rear edge of the tool and the rotating axis of the workpiece.

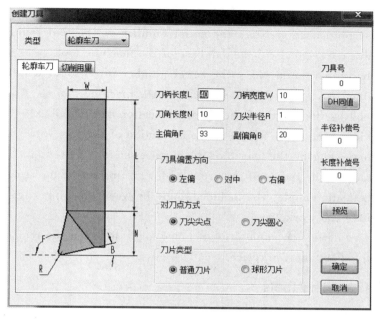

Figure 1-12   Contour turning tool tab

1.4.2 Grooving turning tools

The functional interface of grooving turning tool tab is shown in Figure 1-13, and the specific parameters that need to be configured are as follows.

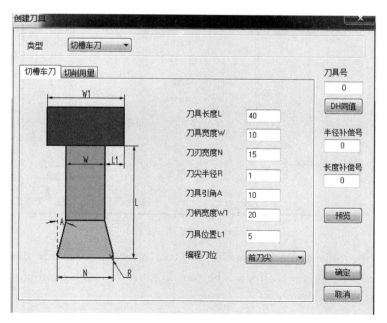

Figure 1-13   Grooving turning tool tab

(1) Tool No.: the serial number of the tool, which is used for the automatic tool change command of post-processing. The tool number is unique and corresponds to the tool magazine of the machine tool.

(2) Radius compensation No.: the serial number of the tool radius compensation value, whose value corresponds to the database of the machine tool.

(3) Length compensation No.: the serial number of the tool length compensation value, whose value corresponds to the database of the machine tool.

(4) Tool length: the length of the clamping section of the tool.

(5) Tool width: the width of the clamping section of the tool.

(6) Tool angle length: the length of the cutting section of the tool.

(7) Nose radius: the radius of the arc used for cutting in the tool tip part.

(8) Main deflection angle: the included angle between the front edge of the tool and the rotation axis of the workpiece.

(9) Secondary deflection angle: the included angle between the rear edge of the tool and the rotating axis of the workpiece.

1.4.3 Thread turning tools

The functional interface of thread turning tool tab is shown in Figure 1-14, and the specific parameters that need to be configured are as follows.

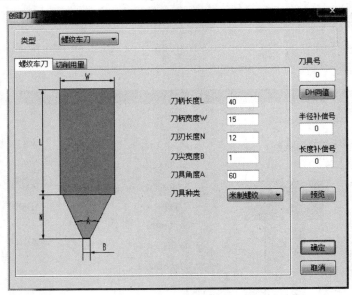

Figure 1-14　Thread turning tool tab

(1) Tool No.: the serial number of the tool, which is used for the automatic tool change command of post-processing. The tool number is unique and corresponds to the tool magazine of the machine tool.

(2) Radius compensation No.: the serial number of the tool radius compensation

value, whose value corresponds to the database of the machine tool.

(3) Tool shank length: the length of the clamping section of the tool.

(4) Tool shank width: the width of the clamping section of the tool.

(5) Blade length: the length of the top of the main cutting edge of the tool.

(6) Tool nose width: the width of thread root.

(7) Tool angle: the angle between the two sides of the cutting section of the tool and the direction perpendicular to the cutting direction, which determines the tooth shape angle of the turned thread.

1.4.4 Drilling tools

The functional interface of drilling tool tab is shown in Figure 1-15, and the specific parameters that need to be configured are as follows.

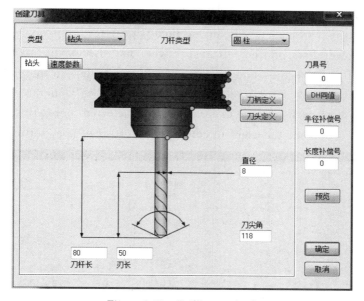

Figure 1-15 Drilling tool tab

(1) Tool No.: the serial number of the tool, which is used for the automatic tool change command of post-processing. The tool number is unique and corresponds to the tool magazine of the machine tool.

(2) Radius compensation No.: the serial number of the tool radius compensation value, whose value corresponds to the database of the machine tool.

(3) Diameter: the diameter of the tool.

(4) Nose angle: the angle of the nose of the drill.

(5) Blade length: the length of the cutter arbor that can be used for cutting.

(6) Cutter arbor length: the distance between the tool nose and the tool shank. The cutter arbor length should be greater than the effective blade length.

1.5　Post-setting of the CNC lathe software

　　Post-setting is to set specific CNC code, CNC program format and parameters for different machine tools and different CNC systems, and generate configuration files. When generating CNC programs, the system generates machining instructions in specific code format required by users according to the definition of the configuration file.

　　Post-setting provides users with a flexible and convenient way to set system configuration. It is of great practical significance to properly configure different machine tools. By setting the system configuration parameters, the CNC program generated by post-processing can be directly input into the CNC machine tool or machining center for processing without modification. If there is no required machine tool among the existing machine tool types, new machine tool types can be added to meet the use requirements, and the new machine tool can be set. The dialog box of post-setting is shown in Figure 1-16. The upper and lower lists on the left respectively list the existing control system and machine tool configuration files, and set the relevant parameters in the middle tabs. In the right test bar, you can select the tool path, and left click the "Generate Code" button. In the code tab, you can see the G code generated by the selected tool path of the current post setting, which is convenient for users to compare the effect of post setting.

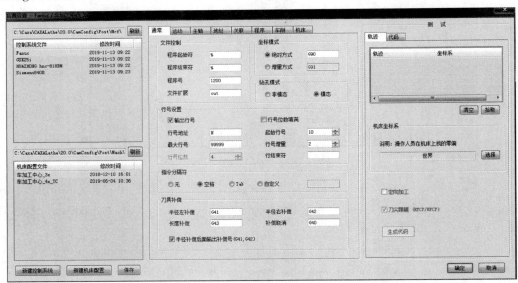

Figure 1-16　Post-setting dialog box of the CNC lathe

Operating instructions:

　　Select CNC lathe→Post-setting, and pop up the post-setting dialog box. Users can add new or change the existing control system and machine tool configuration according to their needs. Press OK to save the user's changes, and press Cancel to discard the changes.

1.5.1 General tab

The basic format of G code can be set in the General tab in the middle section of the post setting dialog box, as shown in Figure 1-17.

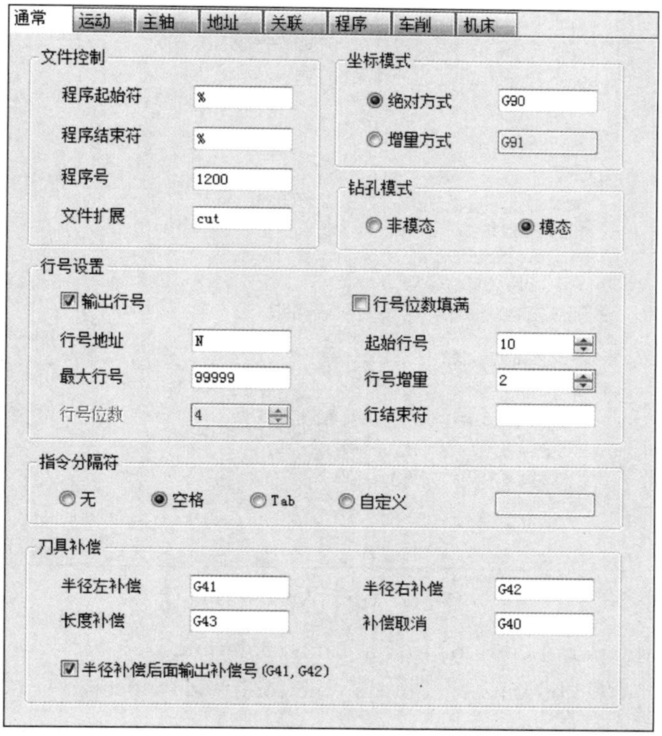

Figure 1-17 The General setting

(1) Document control: set the start and end symbols of G code, set the program number and file extension.

(2) Coordinate mode: set G code instructions in two coordinate modes: absolute coordinate and incremental coordinate relative to the last point.

(3) Line number setting: set whether output the line number or not, the start and end symbols of the line number, the number of digits, whether fill in the number of digits, the address of line number, the maximum and line number and the increment, etc.

(4) Instruction separator: set the separator between CNC instructions.

(5) Tool compensation: Set G code instructions for various tool compensation modes.

1.5.2 Motion tab

In the Motion tab which is in the middle section of the post setting dialog box, you can set the parameters related to tool motion in G code, as shown in Figure 1-18.

(1) Straight line: set G code instructions for tool rapid movement and linear interpolation movement.

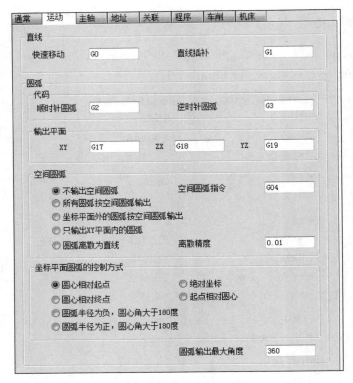

Figure 1-18　Motion tab

(2) Arc: set the parameters of tool arc interpolation.

① Code: set the G code instruction for the tool to do clockwise and counterclockwise arc interpolation.

② Output plane: when setting the plane arc interpolation, G code instructions of different planes where the arc is located.

③ Space arc: sets the processing method of space arc interpolation.

④ Control method of coordinate plane arc: set the meaning of the center point coordinate (I,J,K) in the G code of the arc interpolation segment.

1.5.3　Spindle tab

In the Spindle tab which is in the middle section of the post setting dialog box, you can set the behavior of the machine spindle in the G code, as shown in Figure 1-19.

(1) Spindle: set the G code instructions for spindle forward, reverse and stop rotation.

(2) Speed: set the output mode of the spindle speed.

(3) Coolant: set the G code instructions for switching coolant on and off.

(4) Program code: set the G code instructions for program suspension and termination.

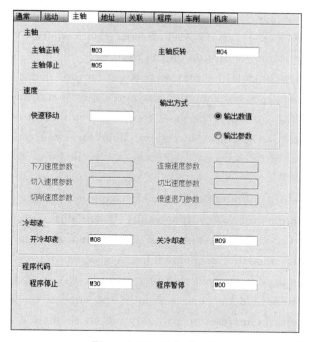

Figure 1-19　Spindle tab

1.5.4　Address tab

In the Address tab which is in the middle section of the post-setting dialog box, you can set the address output format of each instruction of the G code, as shown in Figure 1-20.

Figure 1-20　Address tab

(1) The instruction address list on the left side of the tab lists all available address characters, including X, Y, Z, I, J, K, G, M, F, S, etc. The format of each address character can be modified in the format definition on the right side.

(2) Name: directly controls the address text output in G code. It is usually the same as the address character itself, but sometimes requires special settings. For example, in the G code of the CNC lathe software, the axial coordinate will often output Z, while in the tool path, the axis direction is the X direction. Therefore, you can set the name of the address X to Z, so that in the output G code, the X coordinate of all tool path points will be output in Z.

(3) Modality: before output, the instruction address will determine if the current output value is the same as the last output value. If it is different, the instruction output must be performed in G code. If it is the same, the instruction output will be performed in G code only if OK in the modality drop down list is selected. For example, the modality of the instruction address used for output coordinates, such as X, Y, Z, I, J, K, is set to No. In this way, if the X coordinate of the current point is the same as the previous point, but the Y coordinate is different, the instruction will only output the new Y coordinate.

(4) Coefficient formula: transform the value output by the instruction address. For example, if the formula of the X instruction address is set to "*(−1)", the X coordinates of all tool points will be multiplied by −1 and then output. This item provides a possibility to uniformly modify the output value of G code, but it will affect all the output values of the instruction address in the entire G code, so be careful when using it.

1.5.5 Association tab

In the Association tab which is in the middle section of the post setting dialog box, you can set the instruction address used when outputting various values in G code, as shown in Figure 1-21. The system variable list on the left lists some numerical variables that can modify the instruction address.

1.5.6 Program tab

In the Program tab which is in the middle section of the post-setting dialog box, you can set the G code function of each machining process, as shown in Figure 1-22.

The function name list on the left lists all available function names, and the Function Body tab on the right shows the output format of the selected function.

For example, the LatheLine function is used to output the G code of the linear interpolation machining. The content of the function body is $seq, $speedunit, $sgcode, $cx, $cz, $feed, $eob, @, where the meaning of each variable is as follows:

(1) seq: line number.

(2) speedunit: feed speed unit. Generally, G98 instruction represents feed per minute (mm/min) and G99 instruction represents feed per revolution (mm/r).

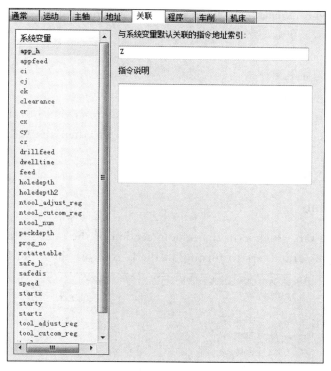

Figure 1-21 Association tab

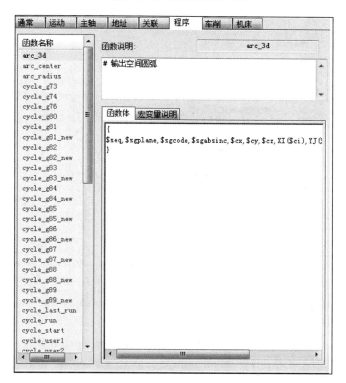

Figure 1-22 Program tab

(3) sgcode: feed command. The linear interpolation command is generally G01 instruction.

(4) cx: radial coordinate value.

(5) cz: axial coordinate value.

(6) feed: feed speed.

(7) eob: terminator, indicating the end of the function.

According to the above definition, if the tool needs to advance to the point (50,20) in the way of linear feed and the feed rate is 20 mm/min, then the G code format of this machining process is:

N10 G98 G01 X50.0 Z20.0 F20;

1.5.7 Turning tab

In the Turning tab which is in the middle section of the post-settings dialog box, you can set some unique parameters to turning in the G code, as shown in Figure 1-23.

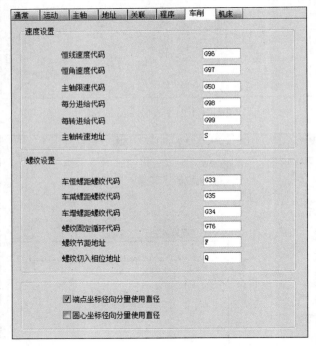

Figure 1-23 Turning tab

(1) Diameter is used for the radial component of the endpoint coordinate. The radial coordinate value in the path uses the radius value, but it is often required to output the diameter value in the G code. When this option is checked, the radial coordinates are output with the diameter value in the G code. For example, the radial coordinate in the tool path is 20, when this option is selected, X40.0 will be output in the G code.

(2) Diameter is used for the radial component of the circle center coordinates. Like linear interpolation, the radius value is also used for the center coordinates of the arc interpolation segment in the tool path. If you need to output the circle center coordinate

with diameter in G code, you can check this option. After checking, if the radial coordinate of the circle center in the tool path is 20, I40.0 will be output in the G code.

1.5.8 Machine tool tab

In the Machine Tool tab which is in the middle section of the post settings dialog box, you can set the information of the machine tool.

As shown in Figure 1-24, the currently selected 3-axis turning machining center can set the initial coordinates, maximum and minimum coordinates of the three linear axes. If the machine tool is 4-axis, you can also set the information about the rotation axis: angle range, rotation axis vector, etc.

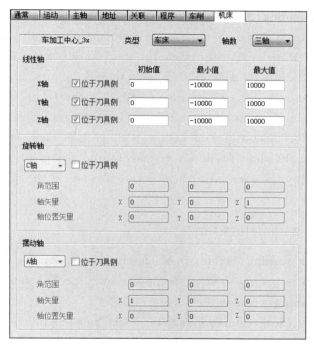

Figure 1-24   Machine tool tab

1.6 Basic instructions

1.6.1 Rough turning(create)dialog box

This function of rough turning is used for rough machining of the outer contour surface, inner contour surface and end face of the workpiece for the rapid removal of excess parts of the blank.

When rough machining the contour, it is necessary to determine the contour to be machined and the blank contour. The contour to be machined is the contour surface of the workpiece after machining, and the blank contour is the contour surface of the blank before machining. The two endpoints of the contour to be machined and the blank contour are connected, and form a closed machining area in which the material will be processed and removed. The contour to be machined and the blank contour cannot be closed or self-intersected separately.

Operation steps:

(1) Select Menu bar→CNC Lathe→Rough Turning, and pops up the rough turning (create) dialog box, as shown in Figure 1-25. In the machining parameter tab, it is first necessary to determine whether the outer contour surface or the inner contour surface or end face is processed, and then the other processing parameters are determined according to the processing requirements.

(2) After determining the parameters, the contour to be processed and the blank contour are picked up. At this time, the contour picking tool provided by the system can be used, and the use of Restricted Chain Pick for the contour composed of multi-segment curves will greatly facilitate the picking. The direction of the pick-up arrow is independent of the actual processing direction when using Chain Pick and Restricted Chain Pick.

(3) Determine the feed and retract points of the cutter. Specifies a point for the location of the cutter before and after machining. Right click to ignore input for that point.

After completing the above steps, a tool path can be generated. Select CNC Lathe→Post-processing, pick up the tool path just generated, and then the machining instructions can be generated.

1.6.2 Machining parameters tab

Left click Rough turning (create)→Machining Parameters to enter the machining parameter tab, which is mainly used to define various process conditions and processing methods in rough turning, as shown in Figure 1-25.

The meaning of each machining parameter is explained as follows.

1. Machining surface type option group

(1) Outer contour: It is machined with an outer contour turning tool, and the default machining direction angle is 180°.

(2) Inner contour: It is machined with an inner contour turning tool, and the default machining direction angle is 180°.

(3) End face: The default machining direction should be perpendicular to the system X axis, that is, the machining angle is $-90°$ or $270°$.

2. Machining parameters option group

(1) Machining angle: It is the angle between the cutting direction and the Z-axis positive direction of the machine tool (X-axis positive direction in software system).

(2) Cutting line spacing: It refers to the cutting depth between lines and the distance between two adjacent cutting lines.

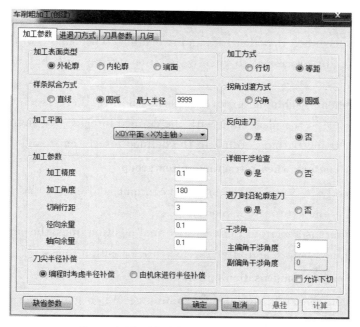

Figure 1-25　Machining parameters tab

(3) Machining allowance: After machining, the remaining amount of the machined surface that has not been machined (compared with the final machining result).

(4) Machining accuracy: The user can control the machining accuracy according to the needs. For straight lines and arcs in the contour, the machine tool can be precisely machined. For the contour composed of spline curves, the system will convert the spline curves into straight segments according to the given accuracy to meet the machining accuracy required by the user.

3. Corner transition mode option group

(1) Circular arc: The tool transits from one side of the contour to the other side in the form of a circular arc when cutting a corner.

(2) Sharp corner: The tool transits from one side of the contour to the other side in the form of a sharp corner when cutting a corner.

4. Spline fitting method option group

(1) Straight line: The spline in the machining contour is fitted with straight line segments according to the given machining accuracy.

(2) Circular arc: The spline in the machining contour is fitted with circular arc segments according to the given machining accuracy.

5. Reverse feeding option group

(1) No: The tool moves in the default direction, that is, the tool moves from the positive direction to the negative direction of the Z axis of the machine to.

(2) Yes: The tool moves in the direction opposite to the default direction.

6. Detailed interference check option group

(1) No: It is assumed that the front and rear interference angles of the tool are 0°, and the groove part is not processed to ensure that the cutting path has no front angle and undercut interference.

(2) Yes: When machining grooves, check whether there is any front angle and undercut interference in the machining with the defined interference angle, and generate an interference-free cutting path according to the defined interference angle.

7. Follow the contour when retracting option group

(1) No: Directly feed and retract at the beginning and end of the tool position line, and do not machine the contour between lines.

(2) Yes: If there is a contour between two tool position lines, the contour between the lines will be machined before and after the next tool position line.

8. Tool nose radius compensation option group

(1) Radius compensation during programming: When generating the tool path, the system will calculate the compensation according to the tool nose radius of the currently used tool (program according to the imaginary tool nose point). The tool nose radius compensation has been considered in the generated code, so it is unnecessary for the machine tool to compensate the tool nose radius.

(2) Radius compensation by the machine tool: When generating the tool path, assume that the tool nose radius is 0mm, and program according to the contour, without calculating the tool nose radius compensation. When the generated code is used for actual machining, the compensation value shall be specified by the machine tool according to the actual tool nose radius.

9. Interference angle option group

(1) Interference angle of main deflection angle: Determine the angle of interference inspection during the front angle interference inspection.

(2) Interference angle of secondary deflection angle: Determine the angle of interference inspection during undercut interference inspection. Available when the "Allow Cutdown" option is checked.

1.6.3 Feed and retract mode

Left click Rough turning(create)→Feed/Retract Mode to enter the feed/retract mode parameter tab. This tab is used to set the feed and retreat mode during rough turning.

1. Feed mode

Relative blank feed mode is used to specify the feed mode when cutting the blank, and relative machining surface feed mode is used to specify the feed mode when cutting the machining surface, as shown in Figure 1-26.

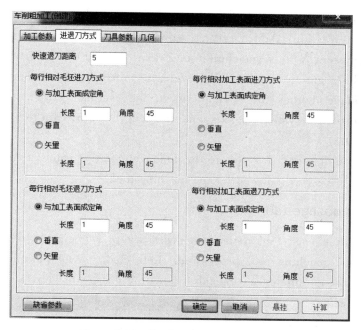

Figure 1-26 Feed and retract mode tab

(1) Fixed angle with machining surface: It refers to adding a feed segment with a certain angle with the cutting direction of the path before each cutting line. The tool vertically feeds to the starting point of the feed segment, and then feeds along the feed segment to the cutting line. The angle defines the included angle between the feed segment and the cutting direction, and the length defines the length of the feed segment.

(2) Vertical: The cutter directly feeding to the starting point of each cutting line.

(3) Vector: Add a feed segment with a certain angle to the positive direction of the system X axis (machine tool Z axis) before every cutting line. The cutter feeds to the starting point of the feed segment, and then feeds along the feed segment to the cutting line. The angle defines the angle between the vector (feed segment) and the positive direction of the system X axis, and the length defines the length of the vector (feed segment).

2. Retract mode

Relative blank retract mode is used to specify the retract mode when cutting the blank, and relative machining surface retract mode is used to specify the retract mode when cutting the machining surface.

(1) Fixed angle with machining surface: It refers to adding a retract segment with a certain angle to the cutting direction of the path before each cutting line. The tool retracts along the segment first, and then retracts vertically from the end point of the segment. The angle defines the included angle between the retract segment and the cutting direction, and the length defines the length of the retract segment.

(2) Vertical: The cutter directly retracting to the starting point of each cutting line.

(3) Vector retraction: Add a retract segment with a certain angle to the positive direction of the system X axis (machine tool Z axis) before every cutting line. The tool retracts along the segment first, and then retracts vertically from the end point of the segment. The angle defines the angle between the vector (retract segment) and the positive direction of the system X axis, and the length defines the length of the vector (retract segment).

Fast retraction distance. It refers to the distance (relative value) that the tool retracts at the maximum feed speed allowed by the machine tool.

1.6.4 Cutting dosage tab

When generating each tool path, it is necessary to set some parameters related to cutting dosage and machine tool processing. Left click "Tool parameter" → "Cutting dosage" to enter the parameter setting tab for cutting dosage, as shown in Figure 1-27.

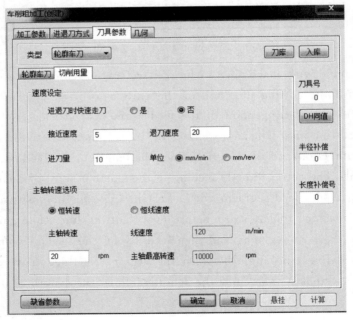

Figure 1-27 Cutting dosage tab

1. Speed setting

(1) Approach speed. The feed speed when the tool approaches the workpiece.

(2) Retract speed. The speed at which the tool leaves the workpiece.

2. Spindle revolution speed option

(1) Constant revolution speed: Keep the spindle revolution speed constant according to the specified spindle speed during cutting until the next command changes the speed.

(2) Constant linear speed: Keep the linear speed constant according to the specified

linear speed value during cutting.

1.6.5 Contour turning tools

Left click Tool Parameters→Contour turning tool to enter the parameter setting tab. This tab is used to set the tool parameters used in machining. For the specific parameter description, please refer to the description in Section 1.4.1.

Task 2 Technological Preparation

1.7 Part drawing analysis

According to the operation requirements of one-way valve, 45 steel is selected as the blank material of this part, and the blanking dimension is set as $\phi 32 \times 40$. Using the excircle of the blank with a diameter of $\phi 32$ as the rough reference, the end face and excircle of the multi-diameter shaft with diameters of $\phi 20$ and $\phi 30$ are roughed, and then finished to ensure the dimensional accuracy and surface quality, and finally cut to ensure the length requirements.

One-way valve is a typical multi-diameter shaft. The two excircles of the multi-diameter shaft need to be processed at the same time in the case of one clamping to ensure better concentricity. If the positioning is divided into two times, the positive difficulty of the part is relatively high, and it is not easy to ensure the working needs of the part.

Note that when clamping the blank, attention should be paid to the extended length of the bar to avoid collision between the cutter and the chuck.

1.8 Technological design

According to the analysis of the part drawing, the technological process is designed as shown in Table 1-2.

Table 1-2 Technological process card

Machining Process Card	Product Model		XSB		Part Number	XSB-01	Page 1	
	Product Name		Suction pump		Part Name	One-way valve	Total 1 page	
Material Grade	C45	Blank Size	$\phi 32 \times 40$	Blank Quality	kg	Quantity	1	
Working Procedure			Work Section	Technical Equipment	Man-hours/min		
No.	Name	Content			Preparation & Conclusion	Single Piece		
5	Preparation	Prepare the material according to the size of $\phi 32 \times 40$	Outsourcing	Sawing machine				

Continued

No.	Working Procedure Name	Working Procedure Content	Work Section	Technical Equipment	Man-hours/min Preparation & Conclusion	Man-hours/min Single Piece
10	Turning	Using the excircle of $\phi32$ as the rough reference, finishing the excircle surface of $\phi20$ and $\phi30$	Turning	Lathe, micrometer	45	30
15	Turning	Reverse positioning with the finished outer circle $\phi20$ and shaft shoulder, and finish machining the end face	Turning			25
20	Cleaning	Clean the workpiece, debur sharp corner	Locksmith			5
25	Inspection	Check the workpiece dimensions	Examination			5

Based on the 10th process and 15th process, namely turning, this training task is designed detailed, and the corresponding working procedure card is formulated as shown in Table 1-3.

Table 1-3 Working procedure card for turning

Machining Working Procedure Card	Product Model	XSB	Part Number	XSB-01	Page 1
	Product Name	Suction pump	Part Name	One-way valve	Total 1 page

Procedure No.		15,20
Procedure name		Turning
material		C45
equipment		CNC lathe
equipment model		CK6150e
fixture		3-jaws chuck
Measuring tool		Vernier caliper
		Micrometer
Preparation & Conclusion time		50min
Single-piece time		40min

Technical requirements:
1. Remove burr and flash.
2. The machined surface of the parts shall be free of scratches and other defects that damage the surface of the parts.
3. Undeclared linear dimension tolerance shall meet the requirements of GB/T 1804—m.

Steps	Content	Cutters	S/ (r/min)	F/ (mm/r)	a_p/ mm	Step hours/min mechanical	Step hours/min auxiliary
1	Workpiece installation						5
2	The outer surface, chamfer, and end face of $\phi10$ and $\phi20$ are roughed, and the finishing allowance is 0.2	Outer circle rough turning tool	1200	0.2	1.5	15	
3	Finishing the outer surface, chamfer, and end face of $\phi10$ and $\phi20$	Outer circle finish turning tool	1500	0.1	0.2	10	
4	Reverse clamping, rough machining $\phi30$ end face	Outer circle rough turning tool	1200	0.2	1.5	10	
5	Finishing $\phi30$ end face and chamfer	Outer circle finish turning tool	1500	0.2	0.2	5	
6	Dismantling and cleaning workpieces						5

Task 3 Software Programming Practice

1.9 Programming practice

1.9.1 Part modelling

The Part to be machined needs to be modeled. The goal of this task is to machine the right end face and the contour, and to establish a closed space between the blank and the processing surface, as shown in Figure 1-28.

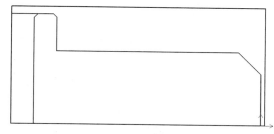

Figure 1-28 Part Modelling

1.9.2 Tool path

1. Roughing tool path

As shown in Figure 1-29, create a contour turning tool and set parameters according to the existing tool. Configure parameters according to the turning process card. For selecting contour curve and blank contour curve, you can select the "Single Selection" in the command bar at the lower left corner.

Note: Contour curve and blank contour curve must be closed.

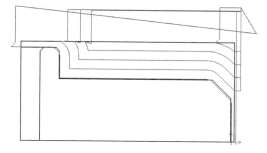

Figure 1-29 Roughing tool path

2. Finishing tool path

For the tool nose radius compensation of the used tool, the radius compensation by the machine tool is selected, as shown in Figure 1-30. The advantage of this method is that the tool radius compensation of the machine tool is directly called, which is convenient for precision adjustment.

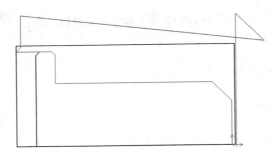

Figure 1-30 Finishing tool path

1.9.3 Reverse clamping for modelling

After reverse clamping, it needs to model the end face and chamfer. This machining is for the end face and chamfer and change the blank contour, as shown in Figure 1-31.

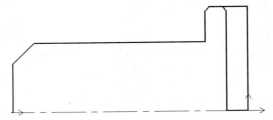

Figure 1-31 Modelling after reverse clamping

1.9.4 Reverse clamping for roughing tool path

1. Reverse clamping for roughing

The tool path for reverse clamping rough machining is shown in Figure 1-32.

2. Finishing tool path after reverse clamping

The finishing tool path after reverse clamping is shown in Figure 1-33.

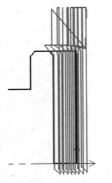 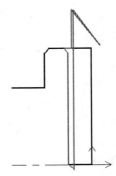

Figure 1-32 Roughing tool path after reverse clamping   Figure 1-33 Finishing tool path after reverse clamping

Project Summary

As a typical machining part of CNC lathes, One-way valves are widely used in various equipment. According to equipment conditions and precision requirements, there will be some differences in the processing technology. Programmers and operators need to formulate the processing technology reasonably according to the processing conditions to improve the processing accuracy of the parts and the production efficiency.

Exercises After Class

1. Fill in the blanks

(1) CNC lathe software provides six drawing methods of circular arc, which are _____, _____, _____, _____, _____ and _____.

(2) The basic application interface of CNC lathe consists of _____, _____, _____, and _____.

(3) Click the _____ button of the mouse, you can activate the menu to determine the position point or pick up the element.

(4) CNC lathe software offers _____ kinds of tool type setting.

(5) If the blank contour is used to define the surface being machined, the specified profile is required to be _____. If the blank contour itself is being processed, the blank contour can also be _____.

2. True or false

(1) When rough turning the contour, the machined contour and that to be machined must be determined. ( )

(2) The machined contour and the blank contour cannot be closed or self-intersected separately. ( )

(3) Constant linear speed instruction is to keep the linear speed constant according to the specified linear speed value during the cutting process. ( )

(4) M09 instruction is coolant on. ( )

(5) The machine coordinate system refers to the X and Z axes established with the machine tool origin as the origin of the coordinate system. ( )

3. Choice questions

(1) In the following instruction, the non-modal code is ( ) instruction.
    A. M03   B. F150   C. S250   D. G04

(2) In the CNC lathe software, the circular arc can be drawn by methods such as ( ).
    A. center + starting point   B. three-point arc
    C. starting point + radius   D. starting point + ending point

(3) The shortcut key that can automatically capture the ends of a line, an arc, a circle,

and a spline is (   ).

  A. M key   B. F key   C. S key   D. space key

(4) The tool magazine management function is used to define and determine the relevant data of the tool, so that the user can obtain tool information from the tool magazine and maintain the tool magazine. This function includes the management of (   ).

  A. contour turning tool   B. thread turning tool and drill tool
  C. grooving turning tool   D. all of the above are included

(5) For drilling, all the machining paths are ultimately on the (   ) axis of the workpiece.

  A. rotation   B. vertical   C. horizontal   D. central

4. Practical operation

(1) Practice CNC lathe software modelling.

(2) Practice the roughing method of CNC lathe.

(3) Practice the finishing method of CNC lathe.

(4) According to the tutorial, try programming reverse clamping machining.

Self-learning test score table is shown in Table 1-4.

Table 1-4 Self-learning Test Score Table

Tasks	Task Requirements	Score	Scoring Rules	Score	Remark
Learn key knowledge points	(1) Understand and master the rough and finish machining of CAXA CNC lathe (2) Understand the comparison between PC programming and manual programming (3) Proficient in using CNC lathe software programming (4) Master the inspection of simulation tool path of CNC lathe (5) Understand post-processing construction settings	50	Understand and master		
Technological preparation	(1) Be able to correctly read the shaft part drawings (2) Be able to determine the process according to the part drawing analysis (3) Be able to correctly program and generate tool path according to the machining process	50	Understand and master		

Ideological and Political Classroom

Project 2 Programming and Machining Training for Pump Body Turning

➤ Mind map

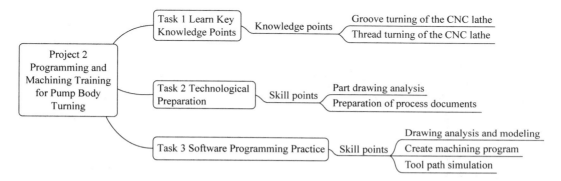

➤ Learning objectives

Knowledge objectives

(1) Understand the groove turning of CNC lathe.

(2) Understand the thread turning of CNC lathe.

Ability objectives

(1) Be able to independently determine the machining process and correctly fill in the process documents.

(2) Be able to correctly create tools and adjust machining parameters according to machining conditions.

(3) Be able to select corresponding inspection tools according to the structural characteristics of parts.

Literacy goals

(1) Cultivate students' enthusiasm for learning.

(2) Cultivate students' hands-on ability.

(3) Establish students' independent thinking ability.

➤ Task introduction

The pump body mainly play the role of connecting and constraining the degree of freedom. The inner hole wall needs to meet the surface roughness requirements to minimize sliding friction and ensure that water does not flow away from the gap. The thread plays the role of connection and sealing, and it is necessary to select appropriate inspection tools for inspection. Since the outside is in contact with human hands, the surface is usually required to be smooth and free of burrs to prevent unnecessary personal injury.

According to the requirements of the part drawing shown in Figure 2-1, develop the processing technology, compile the CNC machining program, and complete the machining of the pump body parts. The blank material of the part is 45 steel, and the surface is required to be smooth.

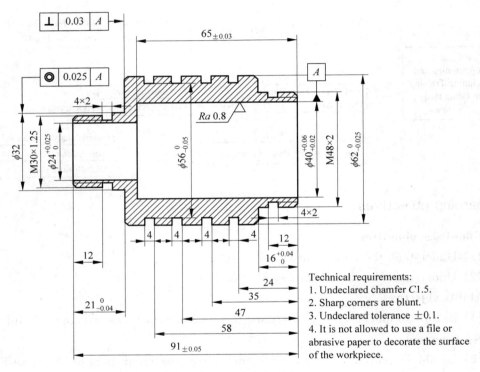

Figure 2-1 The part drawing of pump body

Task 1   Learn Key Knowledge Points

2.1 Groove turning(create)dialog box

This function of groove turning is used to cut grooves on the outer contour surface, inner contour surface and end face of the workpiece.

During groove cutting, the machined contour shall be determined. The machined contour is the workpiece surface contour after machining, and it cannot be closed or self-intersected.

Operation steps:

(1) Select menu bar → CNC lathe → Turning Groove Machining, and pop up the processing parameter tab, as shown in Figure 2-2. In the machining parameter tab, first determine whether the outer contour surface, inner contour surface or end face to be processed, and then determine other processing parameters according to the processing requirements.

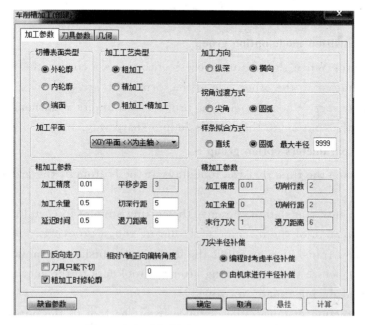

Figure 2-2　Machining parameters tab

(2) After determining the parameters, the contour to be processed can be picked up, and the contour picking tool provided by the system can be used.

(3) After selecting the contour, determine the feed and retract point of the tool. Specifies a point for the tool location before and after machining. Right click to ignore the input of this point.

After completing the above steps, the grooving path can be generated. Left click CNC lathe → Post-Processing, pick up the newly generated tool path, and then generate the machining instructions.

2.1.1　Machining parameters tab

The machining parameters mainly define various process conditions and processing methods in grooving.

The meaning of each machining parameter is explained as follows.

1. Grooved face type option group

(1) Outer contour: The outer contour is grooved, or the outer contour is machined with a grooving tool.

(2) Inner contour: Groove the inner contour, or machine the inner contour with a grooving tool.

(3) End face: Groove the end face, or use a grooving tool to machine the end face.

2. Processing technology type option group

(1) Roughing: Only rough machining the groove.

(2) Finishing: Only finish machining the groove.

(3) Roughing and Finishing: Roughing the groove is followed by finishing.

3. Corner transition mode option group

(1) Round arc: When the cutting process encounters a corner, the tool transits from one side of the contour to the other side in the form of an arc.

(2) Sharp corner: When the cutting process encounters a corner, the tool transits from one side of the contour to the other side in a sharp way.

4. Roughing parameters option group

(1) Delay time: the time that the tool stays at the bottom of the groove during rough grooving.

(2) Cutting depth translation amount: the cutting amount of each longitudinal cutting of the tool (X direction of the machine tool) during rough grooving.

(3) Horizontal translation amount: the horizontal translation amount (Z direction of the machine tool) before the next cutting after the tool cuts to the specified cutting depth translation amount during rough grooving.

(4) Tool retraction distance: the distance from the tool to the outside of the groove before the next line of cutting in the rough grooving.

(5) Machining allowance: the reserved amount of the unprocessed part of the machined surface during roughing groove.

5. Finishing parameters option group

(1) Cutting line spacing: the distance between finishing lines during finishing groove.

(2) Number of cutting lines: the number of machining lines of the tool path during finishing, excluding the repetition number of the last line.

(3) Tool retraction distance: the tool retraction distance before cutting the next line after finishing one line.

(4) Machining allowance: the reserved amount of the unprocessed part of the machined surface during finishing.

(5) Cutting times of the last line: when finishing the groove, in order to improve the surface quality of the processing, the last line is often turned several times under the same feed rate. This defines the number of multiple cuts for the last line.

2.1.2 Cutting dosage tab

The setting of cutting parameters for turning groove processing is shown in Figure 2-3.

For the description of the cutting dosage for groove turning tab, please refer to the instructions in Section 1.6.4.

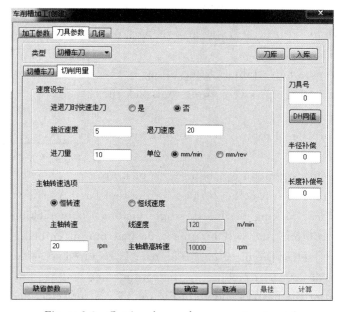

Figure 2-3　Cutting dosage for groove turning tab

2.1.3　Groove turning tools tab

Left click Tool Parameters→parameters setting, as shown in Figure 2-4. This tab is used to set parameters of the grooving turning tool. For specific parameters, please refer to the instructions in Section 1.4.2.

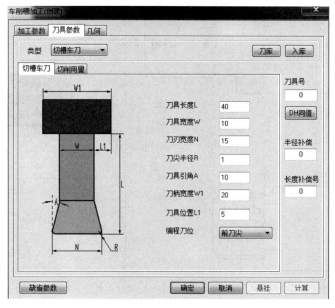

Figure 2-4　Groove turning tools tab

2.1.4 Machining example of turning a groove

The specific steps for machining examples of turning grooves are as follows.

(1) As shown in Figure 2-5, the groove part of the thread tool retraction is the contour to be machined.

(2) Fill in the parameter table. After filling in the parameters in the tab of the groove turning tool, pick the Confirm button.

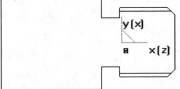

(3) Pick the profile. Prompts user to pick up the contour line.

Figure 2-5 Workpiece drawing

(4) The curve picking menu can be used to pick up the contour line. Press the space key to pop up the tool bar. As shown in Figure 2-6, the tool bar provides three picking methods: single picking, chain picking and limited chain picking. The first contour line will become a dashed line after being picked. Then the system will prompt: select the direction. The user is required to select a direction, which only represents the direction of picking up the contour line and has nothing to do with the machining direction of the tool, as shown in Figure 2-7. After selecting the direction, if the chain picking method is used, the system automatically picks up the contour lines connected from end to end. If the single picking method is used, the system prompts you to continue picking the contour lines. If the limited chain picking method is used, the system continues to prompt you to pick the restriction line. The selected end segment is the left part of the groove, and the groove part turns dashed line, as shown in Figure 2-8.

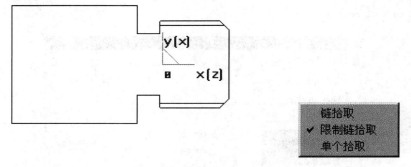

Figure 2-6 Picking tool bar

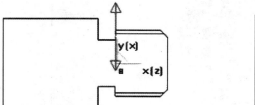

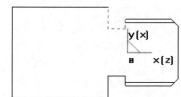

Figure 2-7 Select the contour line direction   Figure 2-8 Limited chain picking method

(5) Determine the feed and retraction point. Specifies a point as the location of the tool before and after machining. Pressing the right mouse button to ignore the input of this point.

(6) Generate tool path. After determining the feed and retraction point, the system generates a tool path, as shown in Figure 2-9.

Note: The contour to be processed cannot be closed or self-intersecting. The generated tool path is closely related to the parameters such as the angle radius and the blade width of the grooving tool. You can draw only the upper part of the tool retraction groove according to actual needs.

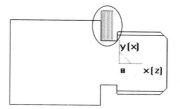

Figure 2-9　Generated tool path

2.2　Thread turning(create)dialog box

This function is a non-fixed cycle way to process threads, which can control various process conditions and processing methods in thread processing more flexibly.

Operation steps:

(1) Left click CNC lathe → Thread turning, and pop up the dialog box of thread turning, as shown in Figure 2-10. The user can determine the processing parameters in the thread parameters tab.

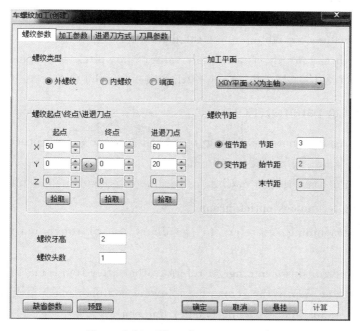

Figure 2-10　Thread parameters tab

(2) Pick the starting point and end point of the thread, as well as the tool feed and retraction points.

(3) After filling in the parameters, click the Confirm button to generate the thread turning tool path.

(4) Left click CNC lathe→Post processing, and pick the tool path just generated, and then the threading instructions can be generated.

2.2.1 Thread parameters tab

Left click the Thread Parameters tag in the dialog box of thread turning (create) to enter the processing parameter tab. Thread Parameters tab mainly contains parameters related to the nature of the thread, such as thread type, pitch, number of heads. The starting point and end point coordinates are derived from the pickup results of the previous step and can be modified by the user.

The meaning of each thread parameter is described as follows:

(1) Starting point coordinate: The coordinate of the starting point of the threading, in mm.

(2) End point coordinate: The coordinate of the end point of the threading, in mm.

(3) Feeding and retraction points: the coordinates of the feeding and retraction points of the threading, in mm.

(4) Thread tooth height: the height of the thread tooth.

(5) Number of thread heads: The number of teeth between the starting point and the end point of the thread.

(6) Thread pitch: there are 5 kinds of thread pitch. Constant pitch is the distance between the corresponding points on two adjacent thread profiles is a constant value; Pitch is constant pitch value; Variable pitch is the distance between the corresponding points on the two adjacent thread profiles is a variable value; Start pitch is the pitch of the thread at the starting end; Final pitch is the pitch of the thread at the termination end.

2.2.2 Machining parameters tab

The Machining Parameters tab (Figure 2-11) is used to set the process conditions and machining methods during thread machining.

The meaning of each thread machining parameter is explained as follows.

1. Machining technology option group

(1) Rough machining: it refers to the direct use of rough machining to process threads.

(2) Rough and finish machining: it refers to that after rough machining according to the specified roughing depth, the remaining allowance (finish machining depth) is cut by fine machining method (such as using smaller line spacing).

2. Parameters option group

(1) Number of the last tool feed: In order to improve the quality of machining surface, the last cutting line sometimes needs to feed repeat with the same feed rate, and

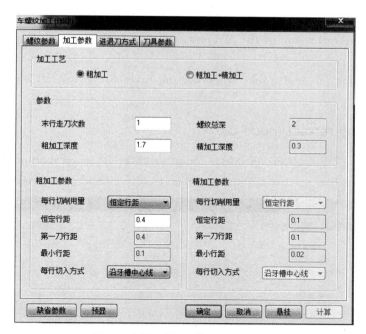

Figure 2-11　Threading parameters

this defines the number of multiple cuts for the last line.

(2) Total thread depth: The total cutting depth for roughing and finishing of the thread.

(3) Roughing depth: The cutting depth of thread for roughing.

(4) Finishing depth: The cutting depth of thread for finishing.

3. Cutting dosage of each line drop-down list

(1) Constant line spacing: the line spacing when the machining is carried out along a constant line spacing.

(2) Constant cutting area: In order to ensure the constant cutting area each time, the cutting depth of each time will be gradually reduced until it is equal to the minimum line spacing. The user needs to specify the first tool line spacing and the minimum line spacing. The cutting depth is specified as follows. The depth of the nth cutting is $\sqrt{n}$ times of the depth of the first cutting.

(3) Variable pitch: The distance between the corresponding points on two adjacent thread profiles is a variable value.

(4) Start pitch: The pitch of the thread at the starting end.

(5) Final pitch: The pitch of the thread at the termination end.

4. Pitching-in method of each line

It refers to the cutting method at the beginning of the thread. The exit mode of the tool at the end of the thread is the same as the cut-in mode.

(1) Along the central line of the alveolar: when cutting in, along the central line of

the alveolar.

(2) Along the right side of the alveolar: when cutting in, along the right side of the alveolar.

(3) Alternating left and right: when cutting in, it alternates left and right along the alveolar.

2.2.3 Feed and retract mode tab

Left click the "Feed/Retract Mode" tag to enter its corresponding parameter tab (Figure 2-12), which is used to set the parameters of feed and retract mode.

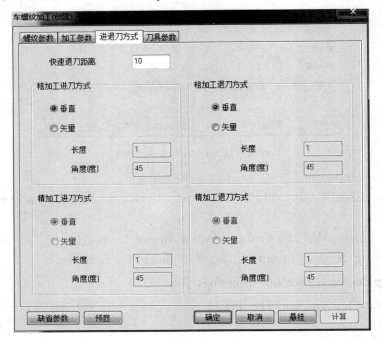

Figure 2-12 Feed and retract mode tab

1. Feed mode

(1) Vertical. It refers to the tool directly feeds to the starting point of each cutting line.

(2) Vector. It refers to adding a feed segment with a certain angle to the positive direction of the system X axis (machine tool Z axis) after every cutting line. The tool feeds to the starting point of the feed segment, and then feeds along the feed segment to the cutting line.

① Length. It defines the length of the vector (feed segment).

② Angle. It defines the included angle between the vector (feed segment) and the positive direction of the system X axis.

2. Retract mode

(1) Vertical. It refers to the tool directly retracts to the starting point of each cutting line.

(2) Vector. It refers to adding a retraction segment with a certain angle to the positive direction of the system X axis (machine tool Z axis) after every cutting line. The tool retracts along the segment first, and then retracts vertically from the end point of the segment.

① Length. It defines the length of the vector (retraction segment).

② Angle. It defines the included angle between the vector (retraction segment) and the positive direction of the system X axis.

3. Fast tool retraction distance

It refers to the distance (relative value) that the tool retracts at the maximum feed speed allowed by the machine tool.

2.2.4 Cutting dosage tab

For the description of the parameters tab of cutting dosage (Figure 2-13), please refer to the instructions in Section 1.6.4.

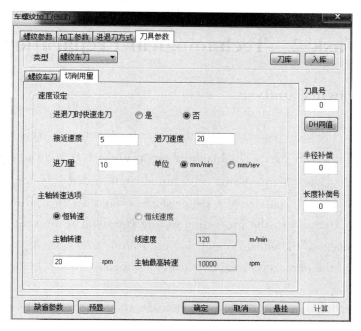

Figure 2-13 Cutting dosage tab

2.2.5 Threading tool tab

Click the Threading Tool tag to enter the threading tool tab (Figure 2-14). This tab is used to set the parameters of threading tool used in processing. For specific parameter description, please refer to the description in Section 1.4.3.

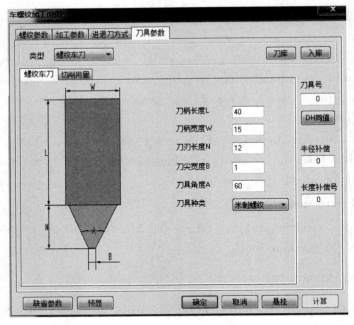

Figure 2-14　Threading tool tab

Task 2　Technological Preparation

2.3　Part drawing analysis

According to the use requirements of the part, 45 steel is selected as the blank material of the pump body part. The blanking size is $\phi 65 \times 100$, and there is a $\phi 20$ through hole in the middle. During processing, $\phi 65$ outer circle is used as the rough reference. Rough and finish the right part to the required size, and then turn around to clamp $\phi 62$ outer circle (pay attention to the protection during clamping to prevent the surface from being pinched), machine the end face and the outer circle of $\phi 32$ and M30 thread stepped shaft on the left of the part to the required size, then turning $M30 \times 1.25$ thread, and check with thread gauge.

Note: When clamping the blank, pay attention to the extended length of the bar to avoid the collision between the tool and the chuck.

2.4　Technological design

According to the analysis of the part drawing, the technological process is designed as shown in Table 2-1.

Table 2-1 Technological process card

Machining Process Card	Product Model	XSB	Part Number	XSB-02	Page 1	
	Product Name	Suction pump	Part Name	Pump body	Total 1 page	
Material Grade	C45	Blank Size	$\phi 65 \times 100$	Blank Quality	kg	Quantity	1

No.	Working Procedure Name	Content	Work Section	Technical Equipment	Man-hours/min Preparation & Conclusion	Single Piece
5	Preparation	Prepare the material according to the size of $\phi 32 \times 40$, and there is a $\phi 20$ through hole in the middle	Outsourcing	Sawing machine		
10	Turning	Using the excircle of $\phi 65$ as the rough reference, finishing $\phi 62$, the major diameter of outer circle and end face of M48 thread	Turning	Lathe, micrometer	130	30
15	Turning	Cutting the tool withdrawal groove of thread and the groove on the circle $\phi 65$		Lathe, Vernier caliper		10
20	Turning	Turning thread		Lathe, Thread gauge		10
25	Turning	Machining the bottom hole of the inner hole and finish boring $\phi 40$ and $\phi 24$ inner hole		Lathe, Inside micrometer		20
30	Turning	Reverse clamping, the outer circle $\phi 62$ is the precision datum, and the major diameter and end face of the thread are processed		Lathe, micrometer		20
35	Turning	Cutting the tool withdrawal groove of thread		Lathe, Vernier caliper		10
40	Turning	Machining M30 thread		Lathe, Thread gauge		10
45	Cleaning	Clean the workpiece, debur sharp corner	Locksmith			5
46	Inspection	Check the workpiece dimensions	Examination			5

Based on turning of the $10^{th} \sim 40^{th}$ process, this training task is designed, and the corresponding working procedure card is formulated as shown in Table 2-2.

Table 2-2 Working procedure card for turning

Machining Working Procedure Card	Product Model	XSB	Part Number	XSB-02	Page 1
	Product Name	Suction pump	Part Name	Pump body	Total 1 page

Procedure No.	10~40	
Procedure name	Turning	
material	C45	
equipment	CNC lathe	
equipment model	CK6150e	
fixture	3-jaws chuck	
Measuring tool	Vernier caliper micrometer Inside micrometer Thread gauge	
Preparation & Conclusion time	125min	
Single-piece time	105min	

Technical requirements:
1. Undeclared chamfer C1.5.
2. Sharp corners are blunt.
3. Undeclared tolerance ±0.1.
4. It is not allowed to use a file or abrasive paper to decorate the surface of the workpiece.

Steps	Content	Cutters	S/(r/min)	F/(mm/r)	a_p/mm	Step hours/min mechanical	auxiliary
1	Workpiece installation						5
2	The outer surface, chamfer, and end face of ϕ62 and M48 are roughed, and the finishing allowance is 0.2mm	Outer circle rough turning tool	1200	0.2	1.5	15	
3	Finishing the major diameter of M48 thread. Finishing ϕ62 out circle and end face	Outer circle finish turning tool	1500	0.1	0.2	10	

Continued

Steps	Content	Cutters	S/(r/min)	F/(mm/r)	a_p/mm	Step hours/min	
						mechanical	auxiliary
4	Cutting the tool withdrawal groove of thread and the groove on the circle $\phi 62$	Slotting tool	600	0.1		15	
5	Rough turning the thread	Thread turning cutter	600	2	0.4	5	
6	Finish turning the thread	Thread turning cutter	600	2	0.1	5	
7	Rough turning the inner hole	Rough turning tool for inner hole	1200	0.2	1.5	10	
8	Finish turning the inner hole	Finish turning tool for inner hole	1500	0.1	0.2	5	
9	Reverse clamping and aligning						10
10	Rough turning the major diameter of M30 thread, the end face, and the chamfer	Outer circle rough turning tool	1200	0.2	1.5	15	
11	Finish turning the major diameter of M30 thread, the end face, and the chamfer	Outer circle finish turning tool	1500	0.1	0.2	10	
12	Grooving the thread tool withdrawal groove	Slotting tool	600	0.1		5	
13	Rough and finish turning the thread	Thread cutter	600	1.5	0.4	10	
14	Dismantling and cleaning workpieces						5

Task 3 Software Programming Practice

2.5 Programming practice

2.5.1 Part model

According to Figure 2-15, establish the enclosed space between the black contour and

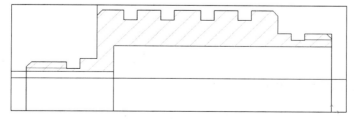

Figure 2-15 Part model

the surface of workpiece, and polish the chamfer on the thread, in order to facilitate cutting and positioning the thread starting point.

2.5.2 Tool path

As shown in Figure 2-16 to Figure 2-19, gradually establish contour turning tools, set parameters based on existing tools, and configure parameters according to the turning process card.

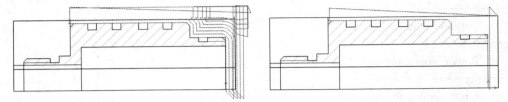

Figure 2-16 The outer circle contour for roughing and finishing

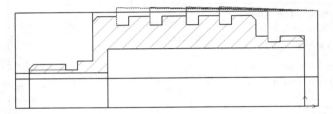

Figure 2-17 Grooving path

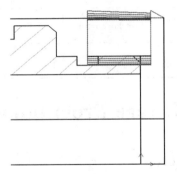

Figure 2-18 Thread turning path

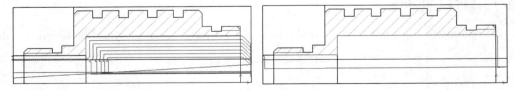

Figure 2-19 Rough and finish turning path for the inner hole

2.5.3 Reverse clamping modelling

As shown in Figure 2-20, after reverse clamping, it needs to model the end face and chamfer. This machining is for the end face and chamfer and make changes to the blank contour.

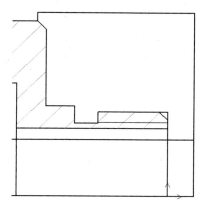

Figure 2-20 Reverse clamping modelling

2.5.4 Reverse clamping machining tool path

Refer to Figure 2-21 to Figure 2-23 for the reverse clamping machining tool path.

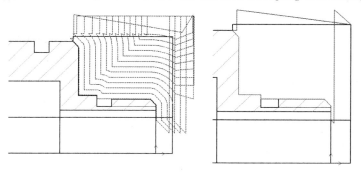

Figure 2-21 Tool path of roughing and finishing M30 thread major diameter and end face after reverse clamping

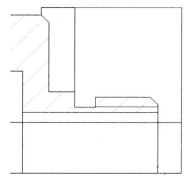

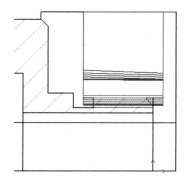

Figure 2-22 Machining thread retract groove tool path after reverse clamping

Figure 2-23 Thread turning tool path after reverse clamping

Project Summary

As a typical processing part of the cavity of CNC lathe, the pump body is widely used in production and life. According to the equipment situation and accuracy requirements, there will be some differences in its processing technology. Programmers and operators need to formulate processing technology reasonably in combination with processing conditions to improve the machining accuracy and production efficiency of parts.

Exercises After Class

1. Fill in the blanks

(1) In the CAXA electronic board, the suffix of the saved drawing file is _____.

(2) An orthogonal line indicates that the drawn line is a line segment _____ to the coordinate axis, either horizontal or vertical.

(3) Turning groove machining can be divided into 3 major steps in CNC lathe software settings, namely _____, _____, _____.

(4) The instructions commonly used for rough and finish machining are _____, _____, _____, _____. Among them, the rough machining instructions are _____, and the finish machining instructions are _____.

(5) In thread parameters, thread pitch includes _____, _____, _____, _____, _____.

2. True or false

(1) The chain pick in CNC lathes does not require the user to specify the starting curve and the direction of the chain search. （　）

(2) Due to the advanced nature of CNC machine tools, any part is suitable for machining on CNC machine tools. （　）

(3) CNC lathe generally specify G86 instruction as the thread turning cycle instruction. （　）

(4) K code in the fixed cycle function instruction refers to the number of repeated processing, which is generally used in incremental mode. （　）

(5) In order to ensure that the cutting area of each cutting is constant, the depth of each cutting is gradually reduced until it is equal to the minimum line spacing. （　）

3. Choice questions

(1) You can use the curve picking tool menu to pick the contour line. Press the space key to pop up the tool menu. Among the following options, （　） is not one of the three picking methods provided by the tool menu.

    A. single pick   B. chain pick

C. restricted chain pick   D. random pick

(2) When the first contour line is picked, it becomes a (   ).

   A. red dashed line   B. red solid line
   C. blue dashed line   D. blue solid line

(3) Which of the following is wrong in the precautions for generating tool path? (   )

   A. The contour to be machined cannot be closed or self-intersecting.
   B. The generated path is closely related to the parameters such as the radius of the grooving tool angle, the width of the cutting edge, etc.
   C. Only the upper part of the retract groove is drawn according to actual needs.
   D. All of the above are error.

(4) The point that do not need to be determined in the thread turning is (   ).

   A. thread starting point   B. thread ending point
   C. thread midpoint   D. feed and retract point

(5) The depth of each cutting in the threading process will be gradually reduced, and the depth is specified as follows: the depth of the nth cutting is (   ) times of that of the first cutting.

   A. $n/2$ B. $\sqrt{n}$ C. $\sqrt[3]{n}$ D. $\sqrt[4]{n}$

4. Short answer questions

(1) Briefly describe how to adjust the machining accuracy of computer programming.

(2) Briefly describe the methods and steps to machine the left-handed threads in thread turning.

(3) What types of parts can be machined in groove turning?

Self-learning test score table is shown in table 2-3.

Table 2-3 Self-learning Test Score Table

Tasks	Task Requirements	Score	Scoring Rules	Score	Remark
Learn key knowledge points	(1) Master processing skills of groove turning (2) Master the use of the tip arc radius compensation instructions (3) Understand threading parameter settings	20	Understand and master		

Continued

Tasks	Task Requirements	Score	Scoring Rules	Score	Remark
Technological preparation	(1) Ability to read part drawings correctly (2) Ability to independently determine the process route and fill in the process documents correctly (3) Be able to write the correct processing program according to the processing process	30	Understand and master		
Programming practice	(1) The programming steps are complete and without missing (2) Generate the tool path without errors (3) The parameter matching is reasonable	50	(1) Understand and master (2) Operation process		

Ideological and Political Classroom

Project 3 Programming and Machining Training for Crank Turning

> Mind map

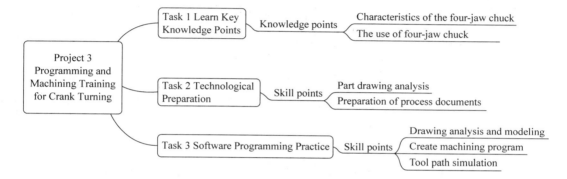

> Learning objectives

Knowledge objectives

(1) Understand the machining accuracy of shaft parts and reasonably arrange the process.

(2) Combine theory with practice and flexibly use knowledge to solve problems.

Ability objectives

(1) Be able to independently determine the machining process and correctly fill in the process documents.

(2) Be able to correctly judge whether the machining process is reasonable.

(3) Be able to select appropriate processing methods and schemes according to the structural characteristics and accuracy of parts.

Literacy goals

(1) Cultivate students' enthusiasm for learning.

(2) Cultivate students' hands-on ability.

(3) Establish students' independent thinking ability.

> Task introduction

Crank parts are widely used in production and life, mainly composed of stepped shaft part and eccentric shaft part. The eccentric shaft part is used for center-change motion.

According to the requirements of the crank part drawing shown in Figure 3-1, develop the machining technology, compile the CNC machining program, and complete the processing of the crank parts. The blank material of the part is 45 steel and the surface is required to be smooth.

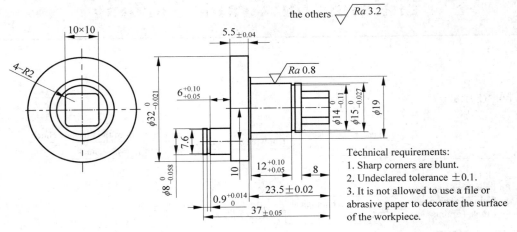

Figure 3-1 The part drawing of crank

Task 1 Learn Key Knowledge Points

3.1 Introduction to the four-jaw chuck

3.1.1 Introduction

  The full name of four-jaw single-action chuck is the manual four-jaw single-action chuck for machine tools, which is composed of a disc body, four lead screws and a pair of jaws. When working, four lead screws are used to drive four jaws respectively, so the common four-jaw single-action chuck has no self-centering function.

  With the increasing demand for work efficiency, hydraulic chuck gradually replaces manual chuck and has been widely used. In order to meet the processing requirements of different workpieces, such as workpieces with rectangular surface, cylindrical blank surface and other irregular surfaces, and some workpiece with eccentricity between the clamping surface and the machining surface, four-jaw hydraulic chuck is usually required. The structure of four-jaw hydraulic chuck produced at home and abroad is that one oil cylinder drives four clamping jaws to clamp at the same time. It is difficult for the workpiece center to be coaxial with the spindle rotation center, and each adjustment of the clamping center must be realized by self-turning clamping jaws, which is very inconvenient. Even if the adjustment is relatively accurate, due to the deviation of the clamping surface of the workpiece itself, according to the principle of three-point centering, it is still impossible to achieve four-point concentric circle, and the fourth jaw is virtual clamp after the three jaws clamping. When the cutting force is large, the clamping force may not be enough to cause workpiece damage or even accidents, which restricts the application of four-jaw hydraulic chuck.

The full name of four-jaw self-centering chuck is a manual four-jaw self-centering chuck for machine tools, which is composed of a disc body, four small bevel teeth and a pair of jaws. Four small bevel teeth mesh with the disc thread. The back of the disc thread has a plane thread structure, and the jaws are installed on the plane thread equally. When the small bevel gear is pulled with a wrench, the disc thread will turn, and the flat thread on its back will make the jaws close to or withdraw from the center at the same time. Because the pitch of the plane rectangular thread on the disc thread is equal, the movement distance of the four jaws is equal, which has the function of self-centering.

There are two kinds of jaws for four-jaw self-centering chuck: the integral jaw and the separating jaw. The integral jaw is a jaw that integrates the base jaw and the top jaw. A pair of integral jaws is divided into four positive jaws and four negative jaws. However, a pair of separating jaws has only four jaws. Each jaw is composed of base jaw and top jaw. Through the transformation of top jaw, it can achieve the function of positive jaw and negative jaw. In addition, soft grippers can also be provided according to user requirements, which can obtain high centering accuracy after being randomly equipped (ground) to meet clamping requirements.

The four-jaw single-action chuck has only one integral jaw. One jaw can be moved independently for clamping eccentric parts and irregularly shaped parts.

3.1.2 Scope of application

1. Four-jaw self-centering chuck

Function: four-jaw synchronous movement is applicable to clamping square parts, as well as shaft and disc parts.

Applicable machine tools and accessories: ordinary lathes, economical CNC lathes, grinders, milling machines, drilling machines and machine tool accessories—indexing heads, slewing tables, etc.

2. Four-jaw single-action chuck

Function: each jaw can be moved independently, which is suitable for clamping eccentric parts and irregularly shaped parts.

Applicable machine tools and accessories: ordinary lathes, economical CNC lathes, grinders, milling machines, drilling machines and machine tool accessories—indexing heads, slewing tables, etc.

3.1.3 Purpose

Lathe accessories are used to hold round, square, and rectangular workpieces for cutting. The four jaws of this chuck cannot be linked and need to be pulled separately, so it can also be used to hold unilateral and eccentric workpieces.

3.2 Method of turning eccentric workpiece

3.2.1 Clamping with a three-jaw chuck

1. Turning method

The eccentric workpiece with short length can be turned on the three-jaw chuck. Firstly turn the outer circle of the non-eccentric part of the eccentric workpiece, and then place a pre-selected shim between any claw of the chuck and the workpiece. After correcting the busbar and eccentricity, and clamping the workpiece, turning can be carried out.

The shim thickness can be calculated by approximate formula: $X = 1.5e$, in which X is the shim thickness, and e is the eccentricity. To make the calculation more accurate, it is necessary to bring the eccentricity correction value, k, into the approximate formula to calculate and adjust the shim thickness, and the approximate formula is: $X = 1.5e + k$, in which $k \approx 1.5$.

$$\Delta e = e - e_{measure}$$

Where Δe is the measured eccentricity error after trial cutting, $e_{measure}$ is the measured eccentricity after trial cutting.

2. Measurement and inspection of eccentric workpieces

When adjusting and correcting the busbar and eccentricity of the workpiece, the dial indicator with magnetic gauge base is mainly used, as shown in Figure 3-2(b), and the turning can only be carried out until the requirements are met. In order to determine whether the eccentricity meets the requirements, a final inspection is also required after the workpiece is turned. The method is to put the workpiece into the V-shaped iron, measure it at the eccentric circle with a dial indicator, slowly rotate the workpiece and observe its runout.

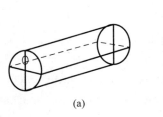
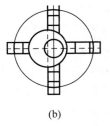

(a)　　　　　　　　　(b)

Figure 3-2　Eccentric adjustment of three-jaw chuck

(a) shape of eccentric workpiece; (b) adjustment of eccentric workpiece

3.2.2 Clamping with a four-jaw single-action chuck

The alignment steps are as follows:

(1) Install the marked workpiece on the four-jaw single-action chuck. During clamping, first adjust the two jaws of the chuck to make them in an asymmetric position, and the other two jaws in a symmetrical position. The eccentric circular line of the workpiece is in the center of the chuck, as shown in Figure 3-3(a).

(2) Put a small flat plate and a needle plate on the bed surface, align the needle tip with the eccentric circle, and correct the eccentric circle. Then align the needle tip with the horizontal line of the outer circle, as shown in Figure 3-3(b), and check whether the horizontal line is horizontal from left to right. Turn the workpiece 90°, check another horizontal line with the same method, then tighten the clamp foot and recheck the clamping condition of the workpiece.

(3) After the workpiece is calibrated, tighten the four jaws again to cut. During the initial cutting, the feed rate should be small and the cutting depth should be shallow. After the workpiece is rounded, the cutting amount can be appropriately increased, otherwise the turning tool will be damaged or the workpiece will be displaced, as shown in Figure 3-3(a).

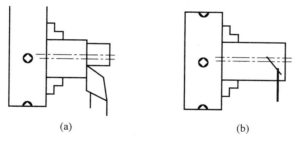

(a)　　　　　　　　　　　(b)

Figure 3-3　Eccentric adjustment of four-jaw single-action chuck

The above two methods are common processing methods, but both have shortcomings. Clamping is troublesome, difficult to align and easy to produce errors. And it is not suitable for mass production. In view of the above shortcomings, an eccentric fixture specially used for mass production is designed. It has been put into production and achieved a certain effect.

3.2.3 Eccentric wheel lathe fixture

The eccentric wheel lathe fixture is mainly used for mass production of eccentric parts. The part drawing of the fixture is shown in Figure 3-4.

When clamping, mark and correct $\phi 60$ inner hole to ensure the tolerance requirements of part shape and position. The clamping method is as follows: clamp the workpiece at the left end of the fixture, after correction by the dial indicator, the right end is clamped on the three-jaw self-centering chuck.

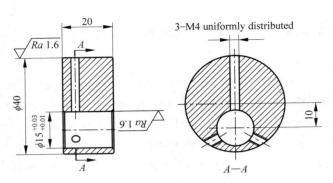

Figure 3-4 The part drawing of the eccentric wheel lathe fixture

Task 2 Technological Preparation

3.3 Part drawing analysis

According to the use requirements of the part, 45 steel is selected as the blank material of the crank part, and the blanking size is $\phi 35 \times 40$. During processing, take the $\phi 35$ outer circle of the blank as the rough benchmark, rough and finish the right end face and the cylinder surface, $\phi 14$ and $\phi 15$, to the required size, grooving, and then turn around, clamp the $\phi 15$ outer circle with four jaws (pay attention to the protection during clamping to prevent the surface from being pinched), machine the left eccentric shaft $\phi 8$ of the part to the required size, and grooving.

Note that when turning the right outer circle $\phi 32$, the turning length should be sufficient. In addition, when clamping the blank, pay attention to the extended length of the bar to avoid the collision between the tool and the chuck.

3.4 Technological design

According to the analysis of the part drawing, the technological process is designed as shown in Table 3-1.

Table 3-1 Technological process card

Machining Process Card		Product Model		XSB		Part Number	XSB-03	Page 1	
		Product Name		Suction pump		Part Name	Crank	Total 1 page	
Material Grade		C45	Blank Size	$\phi 35 \times 40$	Blank Quality	kg	Quantity	1	
Working Procedure						Work Section	Technical Equipment	Man-hours/min	
No.	Name	Content						Preparation & Conclusion	Single Piece
5	Preparation	Prepare the material according to the size of $\phi 35 \times 40$				Outsourcing	Sawing machine		

Project 3 Programming and Machining Training for Crank Turning

Continued

| No. | Working Procedure ||| Work Section | Technical Equipment | Man-hours/min ||
	Name	Content				Preparation & Conclusion	Single Piece
10	Turning	Using the outer circle $\phi 35$ as the rough reference, finishing $\phi 15$ and $\phi 32$ outer circle and end face	Turning	Lathe, micrometer	100	20
15	Turning	Cut the circlip groove		Lathe, Vernier caliper		10
20	Turning	Take the $\phi 32$ outer circle as the precise reference, reverse clamping and adjust eccentricity		Lathe		20
25	Turning	Finishing the $\phi 8$ outer circle and end face		Lathe, micrometer		20
30	Turning	Cut the circlip groove		Lathe, Vernier caliper		10
35	Cleaning	Clean the workpiece, debur sharp corner	Locksmith			5
40	Inspection	Check the workpiece dimensions	Examination			5

Base on turning of the $10^{th} \sim 30^{th}$ process, this training task is designed, and the corresponding working procedure card is formulated as shown in Table 3-2.

Table 3-2 Working procedure card for turning

Machining Working Procedure Card	Product Model	XSB	Part Number	XSB-03	Page 1
	Product Name	Suction pump	Part Name	Crank	Total 1 page

Procedure No.	10~30
Procedure name	Turning
material	C45
equipment	CNC lathe
equipment model	CK6150e
fixture	4-jaw chuck
Measuring tool	Outside micrometer
	Vernier caliper
Preparation & Conclusion time	90min
Single-piece time	80min

Technical requirements:
1. Sharp corners are blunt.
2. Undeclared tolerance ±0.1.
3. It is not allowed to use a file or abrasive paper to decorate the surface of the workpiece.

Continued

Steps	Content	Cutters	S/(r/min)	F/(mm/r)	a_p/mm	Step hours/min	
						mechanical	auxiliary
1	Workpiece installation						5
2	Roughing $\phi 32$ and $\phi 15$ outer circle surface, the chamfer, and the end face. The finishing allowance is 0.2mm	Outer circle rough turning tool	1200	0.2	1.5	15	
3	Finishing $\phi 32$ and $\phi 15$ outer circle surface, the chamfer, and the end face	Outer circle finish turning tool	1500	0.1	0.2	10	
4	Grooving. Machining the circlip groove	Slotting tool	600	0.1		5	
5	Reverse clamping and adjust eccentricity						20
6	Roughing the $\phi 8$ outer circle and end face	Outer circle rough turning tool	1200	0.2	1.5	15	
7	Finishing the $\phi 8$ outer circle and end face	Outer circle finish turning tool	1500	0.2	0.2	5	
8	Dismantling and cleaning workpieces						5

Task 3 Software Programming Practice

3.5 Programming practice

3.5.1 Drawing modeling

The part model is shown in Figure 3-5.

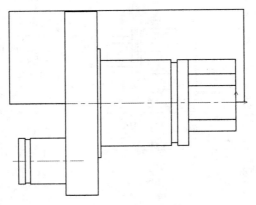

Figure 3-5 Part modeling

3.5.2 Machining tool path

The machining tool path is shown in Figure 3-6 and Figure 3-7.

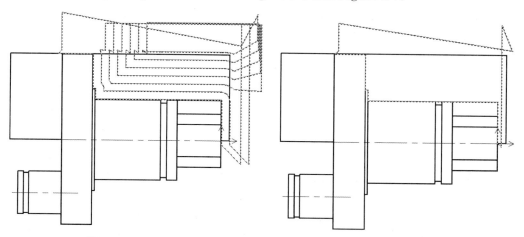

Figure 3-6 Machining tool path of the end face and the outer circle

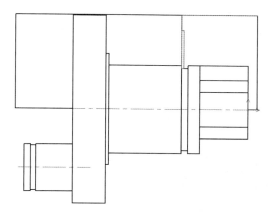

Figure 3-7 Grooving path for the circlip groove

3.5.3 Modeling and machining tool path after reverse clamping and eccentric adjustment

Modeling after reverse clamping and eccentric adjustment is shown in Figure 3-8, and the machining tool path of its outer circle and end face is shown in Figure 3-9. The grooving path for the circlip groove is shown in Figure 3-10.

Note: This crank part has an outer square shape. At present, the lathe cannot process it. It must be milled on the machining center, so there is a margin for the lathe processing.

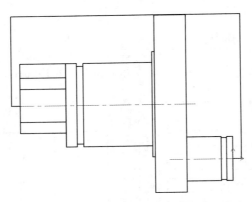

Figure 3-8 Modeling after reverse clamping and eccentric adjustment

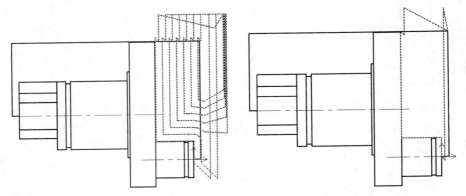

Figure 3-9 Machining tool path of reverse clamping outer circle and end face

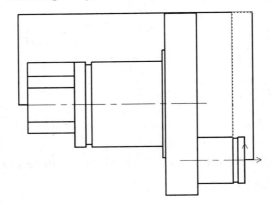

Figure 3-10 Grooving tool path of circlip groove after reverse clamping

Project Summary

As a typical processing part of the cavity of CNC lathe, the crank is widely used in production and life. According to the equipment situation and accuracy requirements, there will be some differences in its processing technology. Programmers and operators need to formulate processing technology reasonably in combination with processing conditions to improve the machining accuracy and production efficiency of parts.

Exercises After Class

1. Fill in the blanks

(1) The full name of four-jaw single-action chuck is the manual four-jaw single-action chuck for machine tools, which is composed of a _____, four _____ and a pair of _____.

(2) There are two kinds of jaws for four-jaw self-centering chuck: _____ and _____.

(3) A four-jaw single-action chuck is suitable for clamping _____ parts and parts.

(4) For machining an eccentric circle, the shim thickness can be calculated by approximate formula: _____.

(5) The eccentric wheel lather fixture is mainly used for _____.

2. True or false

(1) In the application of four-jaw hydraulic chuck, when the cutting force is large, the clamping force may not be enough, resulting in waste products or even accidents, which restricts the application of four-jaw hydraulic chuck. ( )

(2) The four-jaw self-centering chuck is applicable to clamping square parts, as well as shaft and disc parts. ( )

(3) In the approximate formula of shim thickness, e represents the correction value of eccentricity when machining eccentric circle. ( )

(4) The detection method of eccentric workpiece is to put the workpiece into V-shaped iron, measure it at the eccentric circle with a dial indicator, slowly rotate the workpiece and observe its runout. ( )

(5) The commonly used four-jaw fixture is troublesome and easy to produce errors when processing eccentric circles. The eccentric fixture solves these problems well and is suitable for batch production. ( )

3. Choice questions

(1) The three-jaw chuck on the lathe and the gad tongs on the milling machine belong to ( ).

    A. general fixture   B. special fixture
    C. combined fixture   D. accompanying fixture

(2) Combined fixtures are not suitable for ( ).

    A. single-piece and small batch production
    B. machining workpieces with high position accuracy
    C. mass production
    D. trial production of new products

(3) The difference in composition between the four-jaw single-action chuck and the

four-jaw self-centering chuck is ( ).

A. disc body B. clamping device
C. movable jaws D. drive device

(4) In the measurement of eccentric workpiece, there are ( ) steps in the process of clamping and aligning with four-jaw single-action chuck.

A. 1 B. 2 C. 3 D. 4

(5) Which of the following options is not an advantage of the eccentric wheel fixture? ( ).

A. It is difficult to clamp.
B. It is easy to align.
C. It has high accuracy.
D. It is suitable for mass production.

4. Short answer questions

(1) Briefly describe the turning method of eccentric parts.

(2) Briefly describe how to process the 11×11 square.

(3) Practically learn to adjust the eccentric distance with a four-jaw chuck.

(4) Preview the contents of manufacturing engineer software according to the requirements of the crank part drawing (Figure 3-11).

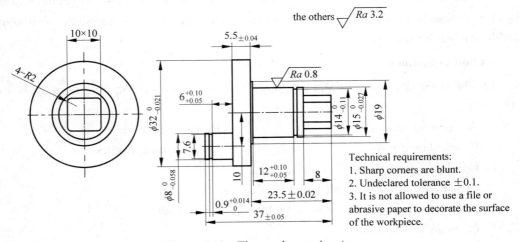

Figure 3-11　The crank part drawing

Self-learning test score table is shown in Table 3-3.

Table 3-3 Self-learning Test Score Table

Tasks	Task Requirements	Score	Scoring Rules	Score	Remark
Learn key knowledge points	(1) Learn about eccentric shaft turning processes (2) Master the eccentric shaft turning method (3) Learn to solve problems of machine complex shaft parts with basic knowledge	20	Understand and master		
Technological preparation	(1) Ability to read part drawings correctly (2) Ability to independently determine the process route and fill in the process documents correctly (3) Be able to write the correct processing program according to the processing process	30	Understand and master		
Programming practice	(1) Be able to complete the programming of complex parts machining (2) Proficient in using machining coordinate system transformation instruction (3) Reasonably configure the machining parameters of the machine tool	50	(1) Understand and master (2) Operation process		

Ideological and Political Classroom

Project 4　Programming and Machining Training for Supporting Seat Milling

➢ Mind map

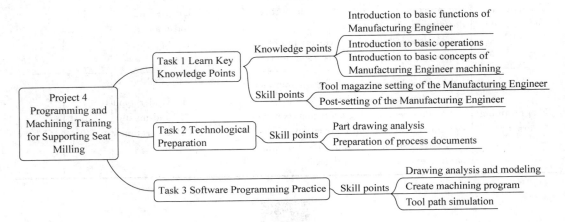

➢ Learning objectives

Knowledge objectives

(1) Understand the operation interface of Manufacturing Engineer 2020 software.

(2) Master the basic drawing and modeling functions of Manufacturing Engineer 2020 software.

(3) Master the function of rough machining of plane area and fine machining of plane contour.

Ability objectives

(1) Master the use of basic preparation instructions and auxiliary instructions.

(2) Master the selection and parameter setting of milling tools.

(3) Be able to independently determine the machining process and correctly fill in the process documents.

(4) Be able to select appropriate processing methods and schemes according to the structural characteristics and accuracy of parts.

Literacy goals

(1) Cultivate students' enthusiasm for learning.

(2) Cultivate students' hands-on ability.

(3) Establish students' independent thinking ability.

Task 1　Learn Key Knowledge Points

4.1　Introduction to basic functions of Manufacturing Engineer software

Manufacturing Engineer 2020 is a newly developed CAD/CAM integrated system based on the solid design platform CAXA 3D 2020. In the aspect of modeling, the precise feature solid modeling technology is adopted, and the wireframe and surface modeling functions of previous versions of manufacturing engineers are inherited and developed. In the aspect of machining, it covers the CNC milling function from two to five axes, and integrates 3D CAD model and CAM machining technology seamlessly. Supports advanced and practical track parameterization and batch processing functions, as well as high-speed cutting. Provides knowledge processing function and general post-processing, also includes many design element libraries.

4.2　Introduction to basic operations

The work interface of Manufacturing Engineer 2020 is shown in Figure 4-1.

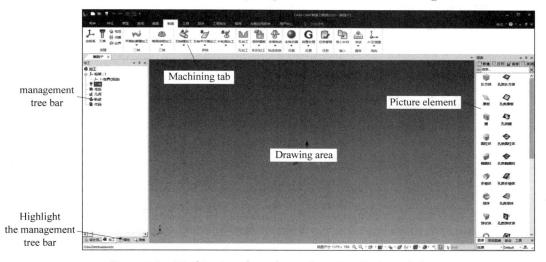

Figure 4-1　Working interface of manufacturing Engineer 2020

4.3　Introduction to basic concepts of Manufacturing Engineer

4.3.1　Modeling

Solid modeling mainly includes extrusion, rotation, guiding, lofting, chamfer, fillet, punching, web-plate, pattern drawing, mold parting, and other feature modeling methods. You can quickly generate 3D solid models from 2D sketch profiles. It provides a variety of functions to build reference planes. Users can build various reference planes according to known conditions.

Surface modeling provides a variety of non-uniform rational B-splines (NURBS) surface modeling methods. Complex surfaces can be generated by scanning, lofting, rotating, guiding, isometry, boundary, grid, etc. It also provides surface line clipping and surface clipping, surface extension, surface stitching according to the average tangent vector or selected surface tangent vector, and splicing between multiple surfaces. In addition, it provides powerful surface blending function, which can realize surface blending methods such as two sides, three sides, and series of faces, as well as equal or variable radius blending.

The system supports the modeling method of mixing solid and complex surface, which is applied to complex part design or mold design. It provides the functions of surface trimming solid, surface thickening to solid, and closed surface filling to solid. In addition, the system also allows the surface of a solid to be generated into a surface for direct reference by users.

The perfect combination of surface and solid modeling methods is a prominent feature of Manufacturing Engineer software in CAD. At each operation step, the prompt area of the software has an operation prompt function. No matter beginners or engineers with rich CAD programming experience, they can quickly grasp the know-how and design their desired part models according to the prompts of the software.

4.3.2 Programming assistant

A new CNC milling programming module is added, which has convenient code editing function, is easy to learn, and is very suitable for manual programming. At the same time, it supports automatic import of code and manually written code, including the track simulation of macro program code, which can effectively verify the correctness of the code. It supports the mutual post-conversion of various system codes to realize the sharing of machining programs on different CNC systems. It has the function of communication transmission. It can realize the code transfer between the CNC system and the programming software through RS 232 port.

Various processing methods such as roughing, semi-finishing, finishing, and clean-up machining

(1) Seven roughing methods are provided: plane area roughing (2D), area roughing, contour roughing, scanning line roughing, cycloid roughing, plunge milling roughing, guide line (2.5 axis) roughing.

(2) Fourteen finishing methods are provided: plane contour, contour guide, surface contour, surface area, surface parameter line, borderline, projection line, contour line,

guide, scan line, limit line, shallow plane, 3D offset, and deep cavity side wall.

(3) Three clean-up machining methods are provided: contour clean-up machining, pen-type back gouging clean-up machining, and regional clean-up machining.

(4) Two kinds of grooving methods are provided: curved groove milling and scanning groove milling.

(5) Multi-axis machining includes four-axis machining and five-axis machining. Four-axis machining includes four-axis curve and four-axis plane cutting; Five-axis machining includes Five-axis isoparametric curves, five-axis side milling, five-axis curve, five-axis curved surface area, five-axis G01 drilling, five-axis orientation, four-axis rotation track. In addition to these machining methods, the system also provides special impeller rough machining and impeller finish machining functions, which can realize the overall machining of impeller and blade.

(6) Macro machining. Provide fillet machining. According to the given plane contour curve, the tool path of machining fillet and machining code with macro instructions are generated. Make full use of the macro program function to make the fillet machining program extremely simple and flexible.

(7) The system supports high-speed machining. Approaching and cutting methods such as oblique cutting and spiral cutting can be set, fillet transition can be set at the corner, and smooth connection can be formed between contours through arc or S-shaped transition to generate smooth tool path, which effectively meets the requirements of high-speed machining for tool path form.

4.3.3 Contour machining

Left click Machining Management in the part feature column, right click in the blank, select Machining→Roughing→Contour Roughing. And pop up the dialog box of Contour Roughing, then Left click Machining Parameter 1, set the machining direction to Down-milling, and check the Z cutting as Layer Height, and enter 0.3 as the cutting height of each layer as the value of layer height. In the XY cutting option, select line spacing, enter 7 as the line spacing, and check ring cutting for the cutting mode, and enter 0.5 as the machining allowance, as shown in Figure 4-2.

Left click the cut in and out tag to enter cut in and out tab, then tick spiral in the mode option group, input the spiral radius of 5 and the pitch of 0.3 in the spiral option group, as shown in Figure 4-3.

Left click the Machining boundary tag and enter machining boundary tab, then tick on the boundary in the option group of tool position in reference to the boundary, as shown in Figure 4-4.

Finally left click Cutting Dosage and set the reasonable machining parameters.

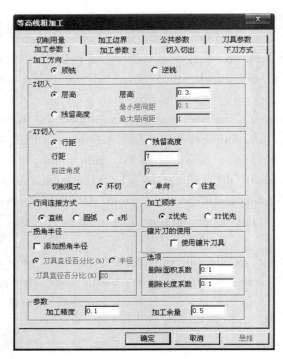

Figure 4-2 Machining parameters tab

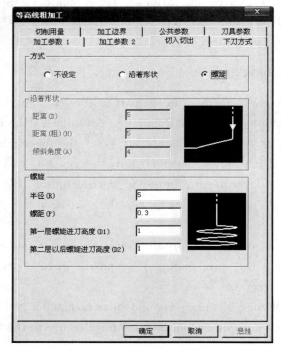

Figure 4-3 Cut in and out tab

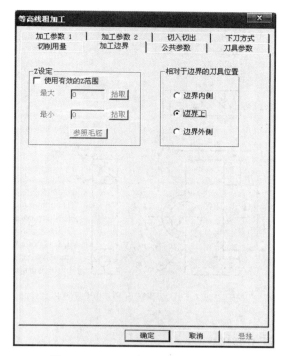

Figure 4-4　Machining boundary tab

Task 2　Technological Preparation

4.4　Part drawing analysis

According to the use requirements of the part, 45 steel can be selected as the blank material of the supporting seat part, and the blank is a square material with a blanking size of 60mm×20mm×40mm.

As shown in Figure 4-5, take the direction perpendicular to 60mm×40mm as the Z axis, process the contour, and then process the precision dimension with a thickness of 12mm through reverse clamping. Finally, clamp the left view direction and machine the cavity to ensure the accuracy and symmetry. Reverse clamping, take the right view as the benchmark to process the countersunk head and through hole to ensure the symmetry.

During milling, the required accuracy of $\phi 15$ hole is high, so the following sequence shall be adopted: roughing → semi-finishing → finishing. The $2\times\phi 5.5$ is a bolt installation hole, so the accuracy is not high, which can be obtained directly through drilling.

Note that this part is in contact with the operator, so after processing, the burr should be removed to ensure that the sharp angle is fully blunted, thus to ensure personal safety during use.

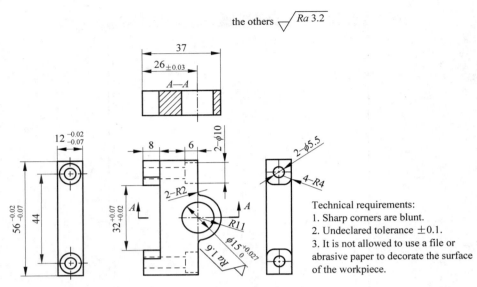

Figure 4-5 Part drawing of supporting seat

4.5 Technological design

According to the analysis of the part drawing, the technological process is designed as shown in Table 4-1.

Table 4-1 Technological process card

Machining Process Card	Product Model	XSB	Part Number	XSB-04	Page 1	
	Product Name	Suction pump	Part Name	Supporting seat	Total 1 page	
Material Grade	C45	Blank Size	60mm×40mm×20mm	Blank Quality	kg	Quantity	1

| | Working Procedure | | | Work Section | Technical Equipment | Man-hours/min | |
No.	Name	Content				Preparation & Conclusion	Single Piece
5	Preparation	Prepare the square material according to the size of 60mm×40mm×20mm		Outsourcing	Sawing machine		
10	Milling	Take the blank surface 60mm×40mm as the rough benchmark, and finish the upper surface		Machining center	Gad tongs	100	20
15	Milling	Finishing the sidewall and the round cavity		Machining center	Gad tongs, Vernier caliper		10

Continued

No.	Working Procedure Name	Working Procedure Content	Work Section	Technical Equipment	Man-hours/min Preparation & Conclusion	Man-hours/min Single Piece
20	Milling	Taking the upper side of 60mm×40mm as the Z-direction datum for positioning, clamping the two sides of 40mm×20mm, reverse clamping, and finishing the plane	Machining center	Gad tongs, Vernier caliper		20
25	Milling	Taking the 60mm×20mm plane as the Z-direction datum, clamp the 60mm×40mm plane, and process the through-hole and countersunk head	Machining center	Gad tongs, Vernier caliper		20
30	Milling	Make Z-direction positioning with two steps and finishing the 32mm×20mm groove	Machining center	Gad tongs, Vernier caliper		10
35	Cleaning	Clean the workpiece, debur sharp corner	Locksmith			5
40	Inspection	Check the workpiece dimensions	Examination			5

Based on the milling of the $10^{th} \sim 30^{th}$ process, this training task is designed, and the corresponding working procedure card is formulated as shown in Table 4-2.

Table 4-2 Working procedure card for milling

Machining Working Procedure Card	Product Model	XSB	Part Number	XSB-04	Page 1
	Product Name	Suction pump	Part Name	Supporting seat	Total 1 page

Procedure No.	10~30
Procedure name	Milling
material	C45
equipment	Machine center
equipment model	AVL650e
fixture	Gad tongs
Measuring tool	Vernier caliper
Preparation & Conclusion time	115min
Single-piece time	110min

Technical requirements:
1. Sharp corners are blunt.
2. Undeclared tolerance ±0.1.
3. It is not allowed to use a file or abrasive paper to decorate the surface of the workpiece.

Continued

Steps	Content	Cutters	S/ (r/min)	F/ (mm/r)	a_p/ mm	a_e/ mm	Step hours/min	
							mechanical	auxiliary
1	Workpiece installation							5
2	Machining the plane 60mm×40mm	ϕ40 face milling cutter	1500	800	1	20	5	
3	Roughing the outer contour with allowance of 0.3mm	ϕ10 end milling cutter	1500	400	3	5	10	
4	Finishing the outer contour	ϕ4 end milling cutter	1500	150	13	1	5	
5	Roughing the round cavity with allowance of 0.1mm	ϕ10 end milling cutter	1500	200	3	5	10	
6	Fine boring the round cavity	ϕ15 boring cutter	400	80	0.5	15	10	
7	Reverse clamping, and take the two sides of 40mm×20mm as the precision datum and the machining plane 60mm×40mm as the Z-direction datum							5
8	Finishing unmachined plane of 60mm×40mm	ϕ40 face milling cutter	1500	400	3	20	10	
9	Take the 60mm×40mm plane as the precision datum, and take the 60mm×20mm plane as the Z-direction datum for positioning							5
10	Drill the centering hole	ϕ8 centering drill	2500	400	2		5	
11	Drill the ϕ5.5 through hole	ϕ5.5 twist drill	600	150	10		10	
12	Milling the countersunk head	ϕ10 end milling cutter	1500	300	6		5	
13	Reverse clamping, taking 60mm×40mm plane as the precision datum and two steps as the Z-axis positioning datum							5
14	Roughing the bottom contour	ϕ10 end milling cutter	1500	600	1	5	10	
15	Finishing the bottom contour	ϕ10 end milling cutter	1500	400			5	
16	Dismantling and cleaning workpieces							5

Task 3 Software Programming Practice

4.6 Programming practice

4.6.1 Create machining coordinate system and blank

First, set up the machining coordinate system, and then create the blank with size of 60mm×40mm×20mm (shown in Figure 4-6). Move the reference point to make the blank cover the workpiece, fill the workpiece and construct the machining part model (shown in Figure 4-7).

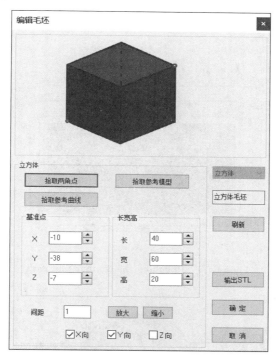

Figure 4-6 Creating a blank

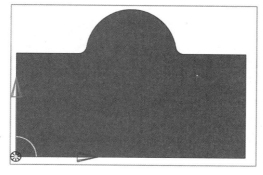

Figure 4-7 Construct machining part model

4.6.2 Machining tool path

Machining tool path is shown in from Figure 4-8 to Figure 4-11.

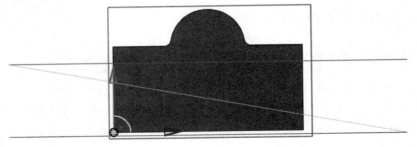

Figure 4-8　Plane machining tool path

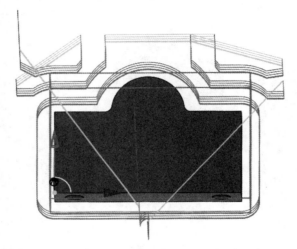

Figure 4-9　Roughing the contour tool path

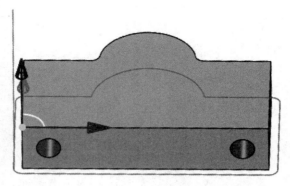

Figure 4-10　Finishing the contour tool path

Note: The processing boundary must be selected when machining the round cavity. This round cavity is the face mated to the crank shaft, and fine boring needs to be selected for machining.

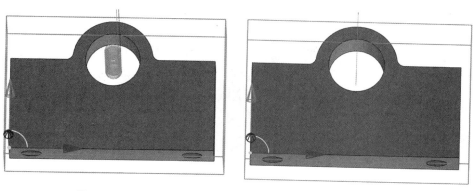

Figure 4-11 Roughing and finishing the round cavity tool path

4.6.3 Machining tool path of reverse clamping

The machining tool path of reverse clamping is shown in Figure 4-12.

Figure 4-12 Roughing and finishing the plane tool path

4.6.4 Hole machining tool path of lateral clamping

The hole machining tool path of lateral clamping is shown in Figure 4-13 and Figure 4-14.

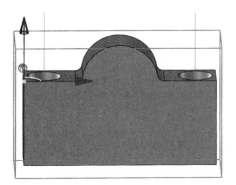 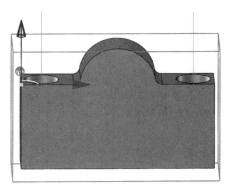

Figure 4-13 Tool path of drilling the hole

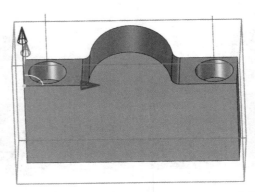

Figure 4-14　Tool path of milling the countersunk head

4.6.5　Machining tool path of bottom irregular groove

The processing for machining of bottom irregular groove includes rough machining and precision machining, as shown in Figure 4-15 and Figure 4-16 respectively.

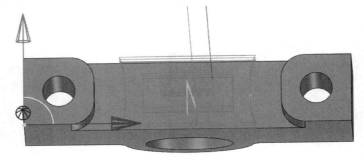

Figure 4-15　Tool path of roughing the bottom irregular groove contour

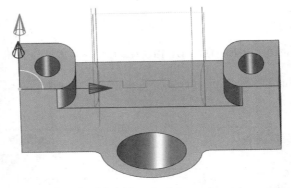

Figure 4-16　Tool path of finishing the bottom irregular groove contour

Project Summary

Through the CNC milling of the supporting seat, master the main CNC machining methods such as plane machining, hole machining, contour machining, etc. Master the machining parameters configuration, tool path inspection and post-processing configuration of the machine tool.

Through task training, cultivate good professional quality, correct programming operation specifications, and basic safety literacy, and develop basic machining quality and safety awareness.

Exercises After Class

1. Fill in the blanks

(1) Manufacturing Engineer 2020 is a newly developed CAD/CAM integrated system based on the platform of _____.

(2) The modeling methods of Manufacturing Engineer 2020 software's solid modeling include: _____, _____, _____, _____, _____, _____.

(3) Manufacturing Engineer 2020 provides surface line clipping and _____, _____, _____ according to _____ or selected surface tangent vector, and splicing between multiple surfaces.

(4) Manufacturing Engineer 2020 supports _____ of various system codes to realize the sharing of machining programs on different CNC systems.

(5) Manufacturing Engineer 2020 provides _____ kinds of finishing methods.

2. True or false

(1) Manufacturing Engineer 2020 can quickly generate 3D solid models from 2D sketch profiles. ( )

(2) Compared with CAXA 3D solid design 2020, Manufacturing Engineer 2020 added a new CNC milling programming module. ( )

(3) Manufacturing Engineer 2020 provides three methods of grooving. ( )

(4) When using the track method to cut grooved parts, for the surfaces on both sides of the groove, one side is down-milling and the other side is up-milling, but the quality of both sides is the same. ( )

(5) After machining, the processing burr shall be removed to ensure that the sharp angle is fully blunted, thus to ensure personal safety during using. ( )

3. Choice questions

(1) Which of the following options is not the work interface column of Manufacturing Engineer 2020? ( )

    A. management tree bar   B. Pixel bar
    C. Machining tabs   D. Drawing brush

(2) Which of the following is not the NURBS surface modeling method of Manufacturing Engineer 2020? ( )

    A. scanning   B. symmetry   C. rotating   D. grid

(3) The processing methods supported by Manufacturing Engineer 2020 include ( ).

    A. Multi-axis machining   B. Macro machining

C. High-speed machining D. All of the above

(4) Common instruction for finishing is ( )instruction.

A. G70   B. G71   C. G73   D. none

(5) For the description of Manufacturing Engineer 2020, which of the following options is wrong? (   ).

A. It Supports advanced and practical track parameterization and batch processing function.

B. It Supports high-speed cutting.

C. It provides knowledge processing function and general post-processing, and contains many design element libraries.

D. Complex operation, small coverage, and limited processing methods.

4. Short answer questions

(1) Briefly describes the use scenarios of plane milling and contour milling.

(2) What are the common tools?

(3) How to select fixtures and tools through process?

Self-learning test score table is shown in Table 4-3.

Table 4-3　Self-learning Test Score Table

Tasks	Task Requirements	Score	Scoring Rules	Score	Remark
Learn key knowledge points	(1) Understand the structure and main parameters of AVL650e vertical machining center (2) Understand the characteristics of milling (3) Be familiar with the classification of commonly used milling cutters and be able to make correct selection of cutters (4) Understand the basic idea of NC programming. (5) Master the method of NC milling programming to process plane, cavity, and hole	25	Understand and master		
Process preparation	(1) Master the correct reading of basic plane part drawings (2) Be able to independently determine the processing process route and correctly fill in the process documents.	25	Understand and master		

Continued

Tasks	Task Requirements	Score	Scoring Rules	Score	Remark
Programming training	(1) Be able to correctly select corresponding equipment and appliances (2) Be able to use software programming correctly and give reasonable parameters	50	(1) Understand and master (2) Operation process		

Ideological and Political Classroom

Project 5　Programming and Machining Training for Base Frame Milling

➢ Mind map

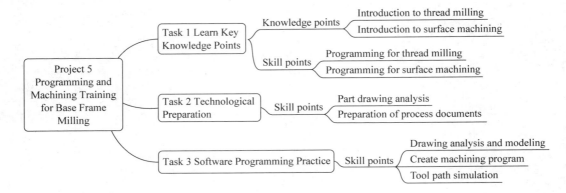

➢ Learning objectives

Knowledge objectives

(1) Understand the three-axis surface machining.

(2) Understand the reason why ball milling cutter cuts curved surface better than end milling cutter.

Ability objectives

(1) Master the use of coordinate system rotation.

(2) Master the instructions of thread machining.

(3) Master the selection and use of threading cutters.

(4) Be able to independently determine the machining process route and correctly fill in the process documents.

Literacy goals

(1) Cultivate students' enthusiasm for learning.

(2) Cultivate students' hands-on ability.

(3) Establish students' independent thinking ability.

Task 1　Learn Key Knowledge Points

5.1　Thread milling

5.1.1　Spiral milling of an inner hole

1. Machining Scope

For blind holes or through holes with larger diameters, spiral milling is often used because twist drilling is too slow or even cannot machine. Moreover, because the tool selected in this way does not have a bottom edge, it is more suitable for machining with small cutting depth, high revolution speed and large feed.

2. Machining characteristic

Spiral milling a hole is a machining method based on the spiral cutting method. When spiral milling a hole, there is a feature: for each spiral milling cycle, the tool moves one cutting height in the Z-axis direction.

3. Explanation

This method is very special in spiral milling inner hole. The essence of its programming is to compile a subprogram with a cutting height as the helix height, and complete the milling of the entire hole by calling the helix subprogram in a cycle. This method is not affected by factors such as milling cutter specifications, so it is widely used in CNC milling machines and machining centers.

5.1.2　Machining thread with single-edge thread milling cutter

1. Machining scope

The same thread milling cutter can mill both left-handed and right-handed threads, both internal and external threads, and is not affected by thread pitch and thread specifications.

2. Machining characteristic

Machining thread with a single-edged thread milling cutter is a machining method based on the spiral cutting method. The principle of thread milling is that the thread milling cutter moves one lead in the direction of the Z-axis (one pitch in case of single line) for each milling cycle.

3. Explanation

This method is very special in thread milling. The essence of its programming is to compile the helix of a lead into a subprogram. By repeatedly calling the helix subprogram, the whole thread milling can be completed. Using this method to machine threads is not affected by parameters such as milling cutter pitch and thread specification, so it is widely used in CNC milling machines and machining centers.

5.1.3 Machining thread with multi-edge thread milling cutter

1. Machining scope

The same thread milling cutter can mill both left-handed and right-handed threads, both internal and external threads. It is mainly used in situations with high production efficiency.

2. Machining characteristic

Each milling cutter has a value, which is the programmed value of the tool fillet radius, that is, the tool radius compensation value for thread milling. For thread milling, the general machining depth can be completed at one time, but if multiple milling is required, it can be completed by modifying the tool radius compensation value.

3. Explanation

This method is very efficient in thread milling, and the programming is also very simple. The essence of its programming is: The thread milling cutter imports (or exports) 1/4 cycle in the XOY plane, and formally processes the thread for 1 cycle; When importing (or exporting) 1/4 cycle in the Z-axis direction, the tool runs 1/4 pitch; When formally processing thread for 1 cycle, the tool runs 1 pitch in the Z-axis direction. By ensuring that each effective tooth on the multi-edge thread milling cutter can participate in milling at the same time, the milling of the whole thread is completed. The key point of this method is to select the pitch of the milling cutter. The corresponding thread milling cutter shall be selected according to the pitch of the thread to be machined. At the same time, this method is affected by factors such as milling cutter pitch and thread specification, but because of its high processing efficiency, it is widely used in CNC milling machines and machining centers.

5.2 Surface machining

Three-axis milling can obtain better curved approximate surfaces. In three-axis milling, use ball-end tools, the linear feed motion in x, y and z directions ensures that the tool can cut to any coordinate point on the workpiece, but the direction of the tool axis cannot be changed. The actual cutting speed at this point on the tool shaft is zero, and the chip removal space at the tool center is also very small.

Surfaces are mathematical entities in CAD models that can accurately represent standard geometric objects (such as planes, cylinders, spheres and torus) and sculpted or free-form geometry. Freeform geometry has numerous applications in the field of design. Surface processing is a multi-purpose application function that fully adapts to the needs and working methods, whether it is machining-area-oriented or operator-oriented. You can define the machining area on the part, and then assign an operation to each area. You can also define the machining process as a series of operations, and each operation has a machining area.

5.2.1 The significance of three-axis surface machining

Three-axis surface machining is a new generation of products that define and manage NC programs. Three-axis surface machining is a three-dimensional geometric machining technology based on three-axis machining technology. It is particularly suitable for mold and tool manufacturers, as well as prototype manufacturers at all branches and industrial levels.

Three-axis surface machining provides a tool path definition for 3-axis manufacturing in workshop that is easy to learn and use. Three-axis surface machining is based on industry-recognized and advanced technology, providing close integration of tool path definition, verification and real-time periodic update.

Thin-walled parts of difficult-to-machine materials with curved surfaces are widely used in industrial production, and shape accuracy is the basic requirement to ensure usability. Because of the low rigidity of thin-walled curved parts, cutting force becomes a sensitive factor of machining deformation. In addition, compared with traditional milling, three-axis high-speed milling has obvious characteristics of small cutting force, which provides an effective method for machining thin-walled curved parts made of titanium alloy and other difficult materials.

Three-axis CNC machining center is widely used for machining because of its simple operation. When machining a surface, the three-axis NC machining center uses interpolation line segments to fit the surface. The quality of the machined surface is affected by the length of the interpolated line segment. A sharp angle is formed at the connection of the straight line. The appearance of sharp corners will lead to the increase of stress concentration.

5.2.2 Three-axis surface machining method

The 3D part geometry and the required surface roughness play a key role in determining the tool path strategy for any given part. The goal of three-axis surface machining is to calculate the path on and along the part surface so that the cutting tool can follow. Generally, the three-axis tool path is projected onto the surface below.

Three-axis NC turning is mainly used to cut the workpiece into the required shape with a tool after rotating the workpiece. When the tool moves on the parallel rotation axis, the inner and outer cylindrical surfaces can be obtained. The formation of the conical surface is the oblique motion of the intersection of the tool and the axis. The formation of the rotating surface is a profiling CNC lathe, and the production of another kind of rotating surface is to control the tool to feed along the curve. It is to use the formed turning tool and feed crosswise. In addition, the machining of thread surface, end plane and eccentric shaft can also be turned with three-axis NC turning.

Three-axis surface cutting mainly includes profile milling, CNC milling or special machining methods. The profile milling must be part prototype. The profiling head of the button-bead is processed under certain pressure to contact with the prototype surface. The motion of the profiling head is converted into inductance, and the machining amplification controls the motion of the three axes of the milling machine, forming a milling cutter combining the button-bead and the profiling radius. The emergence of CNC technology provides more effective means for surface machining.

Task 2　Technological Preparation

5.3　Part drawing analysis

As shown in Figure 5-1, according to the use requirements of the part, 45 steel is selected as the blank material of the base part, and the blank is a square material with size of 95mm×95mm×30mm.

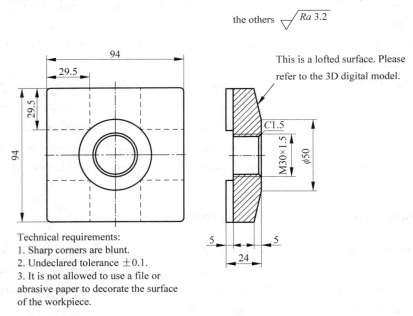

Figure 5-1　Part drawing of base

Part M30 is a key dimension and needs to be matched with other parts. The shape of the part has no matching requirements, and it can be rough machined. Fine tool path is required at the surface.

Note that this part is in contact with the operator, so after processing, the burr should be removed to ensure that the sharp angle is fully blunted, thus to ensure personal safety during use.

5.4 Technological design

According to the analysis of the part drawing, the technological process is designed as shown in Table 5-1.

Table 5-1 Technological process card

Machining Process Card	Product Model		XSB		Part Number	XSB-05	Page 1	
	Product Name		Suction pump		Part Name	Base	Total 1 page	
Material Grade	C45	Blank Size	95mm ×95mm ×30mm	Blank Quality	kg	Quantity	1	
	Working Procedure				Work Section	Technical Equipment	Man-hours/min	
No.	Name		Content				Preparation & Conclusion	Single Piece
5	Preparation	Prepare the square material according to the size of 95mm×95mm×30mm			Outsourcing	Sawing machine		
10	Milling	Taking the blank surface 95mm× 95mm as the rough benchmark, finish the upper surface			Machining center	Gad tongs	160	5
15	Milling	Finishing the curved surface			Machining center	Gad tongs		35
20	Milling	Finishing the sides of 95mm×95mm			Machining center	Gad tongs, Vernier caliper		25
25	Milling	Drill threaded bottom hole and mill thread			Machining center	Gad tongs, Thread gauge		30
30	Milling	Reverse clamping, and mill the plane			Machining center	Gad tongs, Vernier caliper		20
35	Milling	Finishing the cross groove			Machining center	Gad tongs, Vernier caliper		30
40	Cleaning	Clean the workpiece, debur sharp corner			Locksmith			5
45	Inspection	Check the workpiece dimensions			Examination			5

Based on the milling of the $10^{th} \sim 35^{th}$ process, that is milling, this training task is designed, and the corresponding working procedure card is formulated as shown in Table 5-2.

Table 5-2　Working procedure card for milling

Machining Working Procedure Card	Product Model	XSB	Part Number	XSB-05	Page 1
	Product Name	Suction pump	Part Name	Base	Total 1 page

Procedure No.	10～35
Procedure name	Milling
material	C45
equipment	Machine center
equipment model	AVL650e
fixture	Gad tongs
Measuring tool	Vernier caliper
Preparation & Conclusion time	160min
Single-piece time	155min

Technical requirements:
1. Sharp corners are blunt.
2. Undeclared tolerance ±0.1.
3. It is not allowed to use a file or abrasive paper to decorate the surface of the workpiece.

Steps	Content	Cutters	S/ (r/min)	F/ (mm/r)	a_p/ mm	a_e/ mm	Step hours/min	
							mechanical	auxiliary
1	Workpiece installation							5
2	Mill the plane	ϕ40 face milling cutter	1500	400	0.5	20	5	
3	Roughing the curved surface	ϕ40 face milling cutter	1500	800	0.3	20	10	
4	Finishing the curved surface	R5 ball face milling cutter	3000	800	0.1	5	25	
5	Roughing the outer contour	ϕ10 end milling cutter	1500	600	1.5	5	20	
6	Finishing the outer contour	ϕ10 end milling cutter	1500	400	25	5	5	
7	Drill the centering hole	ϕ8 centering drill	2500	200	2		5	
8	Drill the hole	ϕ28.5 twist drill	800	100	30		10	
9	Mill threads	ϕ16 thread milling cutter	800	60			10	
10	Chamfer	ϕ8 centering drill	2500	400	1.5	1.5	5	
11	Reverse clamping, taking the plane 94mm×94mm as the precision datum, and the processing plane in step 2 as the Z-direction datum							5

Continued

Steps	Content	Cutters	$S/$ (r/min)	$F/$ (mm/r)	$a_p/$ mm	$a_e/$ mm	Step hours/min	
							mechanical	auxiliary
12	Roughing the plane	$\phi 40$ face milling cutter	1500	800	2	20	10	
13	Finishing the plane	$\phi 40$ face milling cutter	1500	400	0.3	20	5	
14	Roughing the bottom contour	$\phi 10$ end milling cutter	1500	600	1	5	20	
15	Finishing the bottom contour	$\phi 10$ end milling cutter	1500	400	5	5	10	
16	Dismantling and cleaning workpieces							5

Task 3   Software Programming Practice

5.5   Programming practice

5.5.1   Creating blank and machining coordinate system

Creating blank and machining coordinate system is as shown in Figure 5-2.

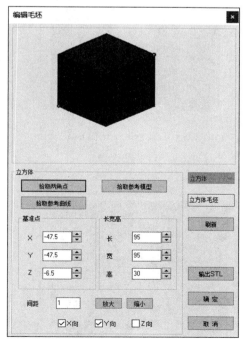
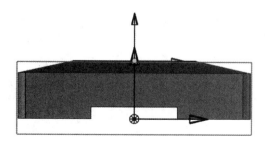

Figure 5-2   Creating blank and machining coordinate system

5.5.2 Machining tool path

The machining tool path is shown in Figure 5-3 to Figure 5-7, which includes plane finishing, surface roughing and surface finishing, outer contour roughing and finishing, as well as thread milling and chamfering. For the plane contour machining here, it is necessary to construct the sketch contour, select the model boundary, and customize the feed and retract points to optimize the tool path. When chamfering the thread hole should pay more attention that the thread bottom hole circle is constructed here.

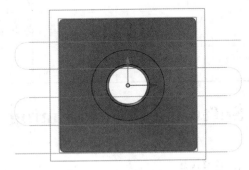

Figure 5-3　Plane finishing tool path

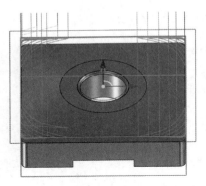 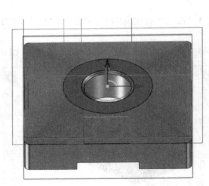

Figure 5-4　Tool path for roughing the curved surface　　　Figure 5-5　Tool path for finishing the curved surface

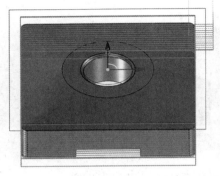 

Figure 5-6　Tool path for roughing and finishing the outer contour

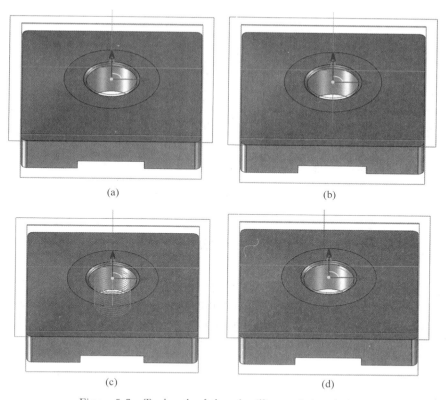

Figure 5-7 Tool path of thread milling and chamfering
(a) Tool path for centering drill; (b) Tool path for Fried Dough Twists drill to drill the thread bottom hole;
(c) Tool path for thread milling; (d) Tool path for chamfering the thread hole

5.5.3 Reverse clamping machining tool path

The reverse clamping machining tool path is shown in Figure 5-8 and Figure 5-9, including reverse clamping plane machining, rough machining of the bottom cross groove, and finish machining.

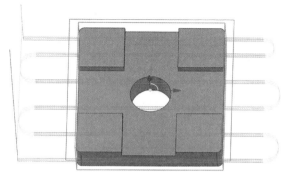

Figure 5-8 Reverse clamping for plane machining

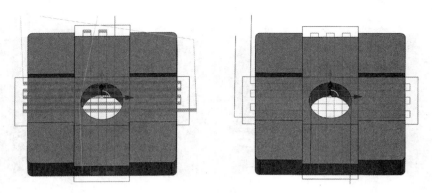

Figure 5-9　Rough and finish machining of bottom cross groove

Project Summary

Through CNC milling of the base, master the selection and setting of CNC milling tools, master the programming of plane machining, cavity machining, thread machining and hole machining, and learn to configure reasonable machining parameters, and check and optimize the tool path. Finally, learn to check code commands when generating code by post-machining.

Through task training, cultivate good professional quality, correct safe operation specifications of machining centers, and basic machining quality awareness.

Exercises After Class

1. Fill in the blanks

(1) Two common tools for spiral milling inner holes are _____ and _____.

(2) In order to obtain a better degree of bending, _____ can be used for machining the curved surface.

(3) Three-axis curved surface machining is a three-dimensional geometric machining technology based on _____.

(4) The common instruction for threaded hole machining is _____.

(5) In the common instruction for threaded hole machining, F code represents _____.

2. True or false

(1) The principle of machining thread with single-edge thread milling cutter is that the thread milling cutter moves one lead in the Z-axis direction for each milling cycle.
(　　)

(2) Multi-edge thread milling cutter has very low efficiency when machining threads, and the programming is also very complex.
(　　)

(3) When machining curved surface with three-axis milling, the actual cutting speed

at this point on the tool axis is zero, and the chip removal space at the tool center is also very small. ( )

(4) Three-axis surface machining is particularly suitable for the requirements of mold and tool manufacturers, as well as prototype manufacturers at all branches and industrial levels. ( )

(5) Compared with traditional milling, three-axis high-speed milling has obvious characteristics of large cutting force. ( )

3. Choice questions

(1) Use FANUC system instruction "G92 X(U) Z(W) F;" to process double thread, then "F" in the instruction refers to ( ).

    A. thread pitch   B. thread lead

    C. feed per minute   D. feed per revolution

(2) Which of the following options is not used for threading? ( )

    A. G32   B. G92   C. G76   D. G81

(3) When using a multi-edge thread milling cutter to process threads, the requirements for tool selection is ( ).

    A. matching thread specification

    B. matching the thread pitch

    C. each effective tooth is simultaneously involved in milling

    D. All of the above

(4) Three-axis curved surface cutting mainly includes ( ).

    A. profile milling   B. CNC milling

    C. special machining method   D. All of the above

(5) Which option is wrong in the following description of three-axis surface machining? ( ).

    A. It is a new generation product that defines and manages CNC programs.

    B. Reliable shape accuracy.

    C. Instant cycle updates.

    D. The operation is complex, the coverage is small, and the machining methods are limited.

4. Short answer questions

(1) Briefly describe what parameters need to be configured when machining the base curved surface?

(2) Please prove that the quality of bevel surface cut by the ball face milling cutter is better than that of end milling cutter.

Self-learning test score table is shown in Table 5-3.

Table 5-3 Self-learning Test Score Table

Tasks	Task Requirements	Score	Scoring Rules	Score	Remark
Learn key knowledge points	(1) Master how to use the coordinate system rotation instructions (2) Master the programming of threads and surfaces	20	Understand and master		
Process preparation	(1) Ability to read part drawings correctly (2) Ability to independently determine the process route and fill in the process documents correctly (3) Be able to write the correct processing program according to the processing process	30	Understand and master		
Programming training	(1) Be able to reasonably select the processing scheme according to the structural characteristics and accuracy of the parts (2) Master the process flow of base milling (3) Can correctly adjust the processing parameters according to the situation of the parts	50	(1) Understand and master (2) Operation process		

Ideological and Political Classroom

Project 6 Programming and Machining Training for Crank Connecting Rod Milling

➤ Mind map

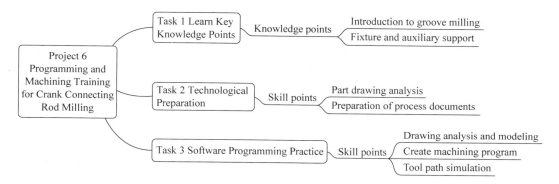

➤ Learning objectives

Knowledge objectives

(1) Understand the groove milling method.

(2) Understand the tooling design and machining of heterosexual parts.

Ability objectives

(1) Master the use of groove milling instruction.

(2) Master the use of tooling fixture to machine irregular parts.

Literacy goals

(1) Cultivate students' enthusiasm for learning.

(2) Cultivate students' hands-on ability.

(3) Establish students' independent thinking ability.

Task 1 Learn Key Knowledge Points

6.1 Groove milling cutters

1. End milling cutter

The end milling cutter (as shown in Figure 6-1) is used to machine grooves and step surfaces, etc. The cutter teeth are on the circumference and end face. Generally, they cannot be fed along the axial direction, but when there are end teeth passing through the center on the end milling cutter, they can be fed axially. The cutting edge has two, three or four edges, and the diameter is $\phi 2 \sim \phi 15$. It is widely used in cut-in milling, high-precision groove machining, etc.

Figure 6-1　End milling cutter

2. Side and face milling cutter

The side and face milling cutter (shown in Figure 6-2) is used to machine various grooves and step faces. The cutter teeth are on the both sides and circumferences. The side and face milling cutter used for cutting the rectangular corners or grooves can be divided into interlocking tooth where cutting edges are interlaced with each other and parallel tooth where cutting edges are arranged in parallel. The parallel tooth side and face milling cutter is the most common tool, while the interlocking tooth side and face milling cutter is used for groove machining of steel. Many indexable tools, including half side and face milling cutter and full side and face milling cutter, are often used in batch production. The full side and face milling cutter is mostly used for groove machining.

Figure 6-2　Side and face milling cutter

3. Metal slitting saw

The metal slitting saw (Shown in Figure 6-3) is used to machine deep grooves and cut off workpieces. There are many cutter tooth on its circumference. In order to reduce the friction during milling, there are secondary declination angles of $15'\sim 1°$ on both sides of the teeth. Features of the metal slitting saw is that the grinding machine can be used to regrind the blade teeth repeatedly. The service life of the grinded metal slitting saw is the same as that of the new metal slitting saw.

4. T-slot milling cutter

The T-slot milling cutter (shown in Figure 6-4) is used to mill the T-slot. The T-slot milling cutter can be divided into taper shank T-slot milling cutter and straight shank T-slot milling cutter. It can be used to process T-slots on various machine tables or other structures. It is a special tool for processing T-slots. It can mill T-slots with required accuracy at one time. The end edge of the milling cutter has an appropriate cutting angle, and the cutter teeth are designed according to the bevel teeth and interlocking teeth, so the cutting is stable and the cutting force is small.

Figure 6-3 Metal slitting saw   Figure 6-4 T-slot milling cutter

6.2 Fixture and auxiliary support

The process of making the workpiece occupying the correct position relative to the tool on the machine tool is called positioning, and the process of overcoming the effect of external force on the workpiece during the cutting process to maintain the accurate position of the workpiece is called chucking, the two kinds of process is called clamping comprehensively. The process equipment is called machine tool fixture when clamping.

Machine tool fixtures can be divided into two categories according to the degree of generalization.

1. Universal fixture

Only a small number of parts need to be adjusted or replaced to clamp different workpieces, such as three-jaw chuck, four-jaw chuck, center, gad tongs, etc.

The structure of universal fixture is complex, which is suitable for mass production, as well as single piece small batch production.

2. Special fixture

The special fixture is specially designed and manufactured for a certain workpiece process. Its structure is simple and compact, and the operation is fast and convenient. The special fixture is applicable to the fixed batch or mass production.

Task 2　Technological Preparation

6.3　Part drawing analysis

As shown in Figure 6-5, according to the use requirements of the part, 45 steel is selected as the blank material of the crank connecting rod part, and the blank is a square material with size of 65mm × 25mm × 25mm. One of the machining schemes is milling shaft and end face of $\phi 8$ precision, finishing the hole of $\phi 8$. Reverse clamping, one clamping and one supporting for the three-jaw chuck, machining $\phi 16$ circle and plane.

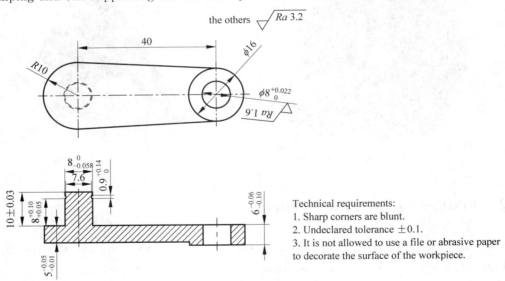

Figure 6-5　Part drawing of the crank connecting rod

Note that this part is in contact with the operator, so after processing, the burr should be removed to ensure that the sharp angle is fully blunted, thus to ensure personal safety during use.

6.4 Technological design

According to the analysis of the part drawing, the technological process is designed as shown in Table 6-1.

Table 6-1 Technological process card

Machining Process Card		Product Model	XSB		Part Number	XSB-06	Page 1	
		Product Name	Suction pump		Part Name	Crank connecting rod	Total 1 page	
Material Grade		C45	Blank Size	65mm× 25mm× 25mm	Blank Quality	kg	Quantity	1
Working Procedure					Work Section	Technical Equipment	Man-hours/min	
No.	Name	Content					Preparation & Conclusion	Single Piece
5	Preparation	Prepare the square material according to the size of 65mm×65mm×25mm			Outsourcing	Sawing machine		
10	Milling	Taking the blank plane 65mm×25mm as the rough benchmark, finish the upper surface			Machining center	Gad tongs	130	5
15	Milling	Finishing the circular dummy club			Machining center	Gad tongs		30
20	Milling	Finishing the outer contour			Machining center	Gad tongs, Vernier caliper		20
25	Milling	Mill the groove			Machining center	Gad tongs		5
30	Milling	Finishing the hole			Machining center	Gad tongs, Vernier caliper		15
35	Milling	Reverse clamping, finishing the plane			Machining center	Gad tongs, Three-jaw chuck		15
40	Milling	Finishing the circular dummy club				Gad tongs, Vernier caliper		20
45	Cleaning	Clean the workpiece, debur sharp corner			Locksmith			5
50	Inspection	Check the workpiece dimensions			Examination			5

Based on the milling of the $10^{th} \sim 40^{th}$ process, that is milling, this training task is designed, and the corresponding working procedure card is formulated as shown in Table 6-2.

Table 6-2 Working procedure card for milling

Machining Working Procedure Card	Product Model	XSB	Part Number	XSB-06	Page 1
	Product Name	Suction pump	Part Name	Crank connecting rod	Total 1 page

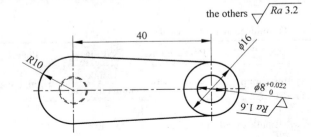

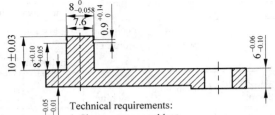

Technical requirements:
1. Sharp corners are blunt.
2. Undeclared tolerance ±0.1.
3. It is not allowed to use a file or abrasive paper to decorate the surface of the workpiece.

Procedure No.	10～35
Procedure name	Milling
material	C45
equipment	Machine center
equipment model	AVL650e
fixture	Gad tongs, Three-jaw chuck
Measuring tool	Vernier caliper
Preparation & Conclusion time	130min
Single-piece time	125min

Steps	Content	Cutters	S/(r/min)	F/(mm/r)	a_p/mm	a_e/mm	Step hours/min mechanical	auxiliary
1	Workpiece installation							5
2	Mill the plane	φ40 face milling cutter	1500	400	0.5	20	5	
3	Roughing the circular dummy club and the plane	φ40 face milling cutter	1500	800	0.3	20	20	
4	Finishing the circular dummy club and the plane	φ10 end milling cutter	1500	600	0.1	5	10	
5	Roughing the outer contour	φ40 face milling cutter	1500	600	2	20	15	
6	Finishing the outer contour	φ10 end milling cutter	1500	400	25	5	5	
7	Machining the circlip groove	φ20×0.9 groove milling cutter	1000	100			5	
8	Drill the centering hole	φ8 centering drill	2500	200	2		5	
9	Drill the hole	φ7.8 twist drill	800	80	8		5	
10	Hole-reaming	φ8 reaming tool	600	40			5	
11	Reverse clamping, three-jaw chuck positioning, gad tongs supporting cantilever							5

Steps	Content	Cutters	S/(r/min)	F/(mm/r)	a_p/mm	a_e/mm	Step hours/min	
							mechanical	auxiliary
12	Roughing the plane		1500	800	2	20	10	
13	Finishing the plane	ϕ40 face milling cutter	1500	400	0.3	20	5	
14	Roughing the circular dummy club and the plane	ϕ40 face milling cutter	1500	600	1	20	10	
15	Finishing the circular dummy club and the plane	ϕ10 end milling cutter	1500	400	1	5	10	
16	Dismantling and cleaning workpieces							5

Task 3 Software Programming Practice

6.5 Programming practice

6.5.1 Creating blank and machining coordinate system

At first creating blank and then create machining coordinate system, as shown in Figure 6-6.

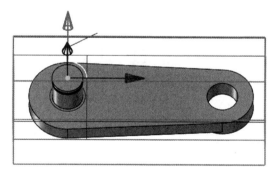

Figure 6-6 Create machining coordinate system

6.5.2 Tool path

The tool path is shown in Figure 6-7 to Figure 6-11, which includes machining the plane, rough and finish machining of circular tables and planes, milling of cirelip grooves, rough and finish machining of outer wheel profiles, and hole machining.

Note: When mill clamp spring grooves, the groove contour circle here is smaller than the diameter of the circular table, and it is necessary to avoid it.

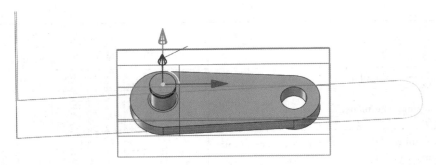

Figure 6-7　Tool path of machining the plane

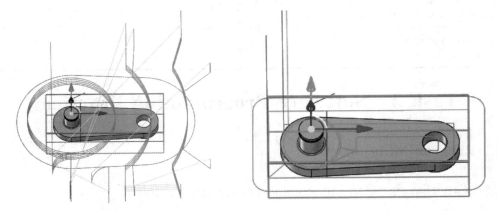

Figure 6-8　Tool path for roughing and finishing the circular table and plane

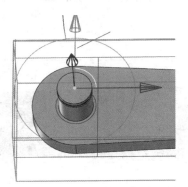

Figure 6-9　Tool path for milling the circlip groove

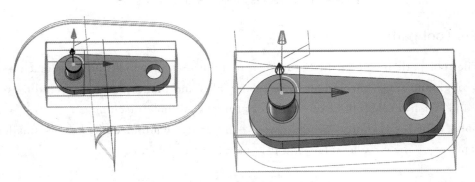

Figure 6-10　Tool path for roughing and finishing the outer contour

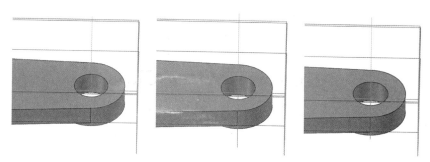

Figure 6-11　Tool path for hole machining

6.5.3　Reverse clamping tool path

The reverse clamping tool path is shown in Figure 6-12 and Figure 6-13, including rough and fine machining of the plane and contour.

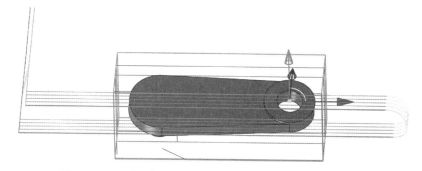

Figure 6-12　Tool path for roughing and finishing the plane

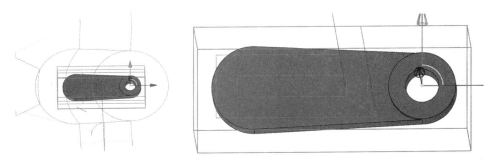

Figure 6-13　Tool path for roughing and finishing the contour

Project Summary

Through the programming and machining of crank connecting rod, master the flexible use of groove milling cutter and fixture.

The processing method is not unique. It is necessary to select the machine tool reasonably according to the existing conditions and drawing accuracy, reduce the clamping times as much as possible, and ensure that the working hours are as short as possible.

Through task training, cultivate good professional quality, correct safe operation specifications of machining centers, and basic machining quality awareness.

Exercises After Class

1. Fill in the blanks

(1) The types of groove milling cutter are _____, _____, _____, _____.

(2) T-slot milling cutters can be divided into _____ and _____.

(3) Machine tool fixtures can be divided into two categories, namely _____ and _____.

(4) Special fixtures have the advantages of _____, _____, _____, etc.

2. True or false

(1) Generally, the end milling cutter can feed along the axial direction. ( )

(2) The parallel tooth side and face milling cutter is the most common tool, while the interlocking tooth side and face milling cutter is used for groove machining of steel. ( )

(3) The metal slitting saw (see Figure 6-3) is used to machine deep grooves and cut off workpieces. There are many cutter teeth on its circumference, and it can be repeatedly sharpened to extend its service life. ( )

(4) The T-slot milling cutter is a special tool for machining T-slots, but its cutting force is too large, which often causes the tool to vibrate. ( )

(5) Compared with the universal fixtures, special fixtures are more suitable for fixed batch or mass production. ( )

3. Short answer questions

Please use another scheme to process the specified parts according to your knowledge.

Self-learning test score table is shown in Table 6-3.

Table 6-3 Self-learning Test Score Table

Tasks	Task Requirements	Score	Scoring Rules	Score	Remark
Learn key knowledge points	(1) Understand the basic composition of parts (2) Master various machining methods (3) Design new processing schemes	25	Understand and master		

Continued

Tasks	Task Requirements	Score	Scoring Rules	Score	Remark
Process preparation	(1) Be able to read the part drawing correctly (2) Be able to independently determine the processing route (3) Be able to write correct program according to the machining process	25	Understand and master		
Programming training	According to the process plan, prepare the program and execute the machining	50	(1) Understand and master (2) Operation process		

Ideological and Political Classroom